STUDENT UNIT G

A2 Biology
UNIT 4

AQA Specification B

Module 4: Energy, Control and Continuity

Keith Hirst

Philip Allan Updates
Market Place
Deddington
Oxfordshire
OX15 0SE

Orders
Bookpoint Ltd, 130 Milton Park, Abingdon, Oxfordshire, OX14 4SB
tel: 01235 827720
fax: 01235 400454
e-mail: uk.orders@bookpoint.co.uk
Lines are open 9.00 a.m.–5.00 p.m., Monday to Saturday, with a 24-hour message answering service. You can also order through the Philip Allan Updates website: www.philipallan.co.uk

ISBN-13: 978-0-86003-476-6
ISBN-10: 0-86003-476-3

This Guide has been written specifically to support students preparing for the AQA Specification B A2 Biology Unit 4 examination. The content has been neither approved nor endorsed by AQA and remains the sole responsibility of the author.

Printed by MPG Books, Bodmin

P00801

Contents

Introduction

Content Guidance

Questions and Answers

Introduction

About this guide

This guide is for students following the AQA Specification B A2 Biology course. It deals with Unit 4, which examines the content of **Module 4: Energy, Control and Continuity**. The key to success is examination technique. You should always have at the back of your mind the type of questions that can be asked, when both learning and revising a topic. This Introduction is devoted to the aims of the specification and to learning and revision skills. The Content Guidance section provides an outline of the topics you need to know and understand and includes detailed explanations of some of the topics. The Question and Answer section contains sample unit test questions, together with candidate responses which are accompanied by examiner's comments.

The best way to use this book is to:

- revise a topic using the Content Guidance section as a guide
- attempt the relevant question(s) *without looking* at the candidate responses
- compare your responses with the candidate responses and examiner's comments and see what marks you might have achieved
- revise the parts of the topic for which you did not obtain high marks

The aims of the A2 specification

A2 biology encourages you to:

- develop knowledge and understanding of concepts of biology
- develop the skills to use this knowledge and understanding in new situations
- develop an understanding of the methods used by scientists
- be aware of advances in technology that are relevant to biology
- recognise the value and responsible use of biology in society
- sustain and develop an interest in, and enjoyment of, biology
- show knowledge and understanding of the facts, principles and concepts from different areas of biology and to make and use connections between them

How the unit test assesses these aims

Very few marks in the unit test are given for simple recall of knowledge. Most of the marks are given for being able to:

- demonstrate understanding of concepts
- apply knowledge and understanding

These two areas include many skills, most or all of which will be addressed later in the unit test. In summary, you should be able to do the following.

- Draw on your knowledge to show understanding of the ethical, social, economic, environmental and technological implications and applications of biology. Read scientific articles in newspapers and periodicals, and watch documentaries on current affairs that deal with scientific issues so that you are aware of different viewpoints on controversial issues.
- Select, organise and present relevant information clearly and logically.
- Practise answering the longer section B questions that require continuous prose.
- Describe, explain and interpret phenomena and effects in terms of biological principles and concepts, presenting arguments and ideas clearly and logically. Make sure you know the difference between the 'trigger' words *explain* and *describe*.
- Interpret (and translate from one form into another) data presented as continuous prose, or in tables, diagrams, drawings and graphs. You will be presented with data in many different forms in the unit test. Make sure that you have practised questions involving comprehension, graphs, tables and diagrams.
- Apply biological principles and concepts in solving problems in unfamiliar situations, including those relating to the ethical, social, economic and technological implications and applications of biology. There will always be unfamiliar data in the unit test and the examiner will ask you to 'suggest' explanations. Again, make sure you have practised many examples of this type of question.
- Assess the validity of biological information, experiments, inferences and statements. Don't leave your experimental skills in the laboratory. Questions which involve interpreting and evaluating data will usually appear in the unit tests.

A unit test will require the use of most, if not all, of these skills.

Weightings

Unit 4 is assessed by a unit test and carries 15% of the total A-level mark. Of this 15%:

- 6.5% of the marks are given for demonstrating knowledge and understanding of the unit content
- 8.5% of the marks are given for being able to apply this knowledge and understanding in new situations

Command terms

Examiners use *trigger words* to advise you which skill they are testing. You must know what the examiner wants when these trigger words appear in a question.

Name/what is the name of...?

This usually requires a technical term or its equivalent. Answers to this type of question normally involve no more than one or two words. Do not waste time by repeating the question in the answer.

List...

This requires you to give a number of features or points, each often no more than a single word, so do not go into further detail.

Define/what is meant by...?

'Define' requires a statement giving the meaning of a particular term or word. 'What is meant by...?' is used frequently in questions on a comprehension passage. It emphasises that a formal definition as such is not required.

Outline...

This means give a brief summary of the main points. There are two good indications as to the amount of detail required. These are the mark allocations and the space allowed for the answer — usually two lines per mark.

Describe...

This means no more than it says: 'Give a description of...'. So 'Describe a curve on a graph' requires a description of the shape of the curve, preferably related to key points or values; 'Describe an experiment' means give an account of how such an experiment might be carried out.

Describe how you...

The emphasis here is on the word you and the expression is often used when asking questions about experimental design. What is required is an account of how something could be done by you as a student working in an ordinary school or college laboratory.

Evaluate...

Evaluating is more than just listing advantages and disadvantages. It requires an explanation. Evaluating the evidence for and against a particular point of view requires an explanation of each of the points being made.

Explain...

This requires you to give a reason or interpretation, not a description. The term 'describe' answers the question 'what?' The term 'explain' answers the question 'why?' Thus, 'Explain a curve on a graph' requires a biological reason for any change of direction or pattern that is evident.

Suggest...

Suggest is used when it is not possible to give the answer directly from the facts you have learned. The answer should be based on your general understanding of biology rather than on recall of learnt material. It also indicates that there may be a number of correct alternatives.

Give the evidence for... /using examples from...

Answers to questions involving these phrases must follow the instructions. Marks are *only* awarded for appropriate references to the information provided in the question.

Plot/sketch...

These terms refer to the drawing of graphs. 'Plot' means that the data should be presented as an appropriate graph on graph paper with the points plotted accurately. 'Sketch' requires a simple estimate of the expected curve, and can be made on ordinary lined paper. However, even in a sketched graph, the axes should be correctly labelled.

Calculate...

This term is used where the only requirement is a numerical answer expressed in appropriate units. The additional instruction, 'Show your working', will be used if details or methods are required. Make sure that you can calculate percentages and proportions, since these appear in most unit tests.

Revision planning

Key words

A biological specification contains so many unfamiliar words it can appear to be a foreign language. It is important that you know the meaning of all of these words so that you know what is being asked in a question and can use the words correctly in your responses.

Below are some extracts from Module 4. The biological words that you need to know are in bold.

- The structure of a **myelinated motor neurone**.
- The establishment of a **resting potential** in terms of **differential membrane permeability** and the presence of **cation pumps**.
- The initiation of an **action potential** and its all-or-nothing nature, explained by changes in membrane permeability leading to **depolarisation**.
- The passage of an **action potential** along **non-myelinated** and myelinated **axons**, resulting in **nerve impulses**.

It is a good idea to go through the specification content listed in Content Guidance, underlining biological words and then writing a definition of each one, for example:

- **Resting potential** — the difference in charge across the outer membrane of a neurone.
- **Action potential** — the change that occurs in the electrical charge across the outer membrane of a neurone during the passage of a nerve impulse.

This will give you the biology vocabulary that is essential both to understand and to answer questions.

Revision progress

You may find it useful to keep track of how your revision is going by drawing the table below, listing the topics in the first column.

Module topic	Revised (N/P/F)	Self-evaluation (1–5)
Energy supply	F	5
Photosynthesis	F	4
Respiration	F	3
Survival and coordination	P	2
Homeostasis	N	
Nervous coordination	N	
Analysis and integration	N	
Muscles as effectors	N	
Inheritance	N	
Variation	N	
Selection and evolution	N	
Classification	N	

Complete column 2 to show how far you have got with your revision.

N = not yet revised
P = partly revised
F = fully revised

Complete column 3 to show how confident you are with the topic.

5 = I am confident I could answer any question on this topic
1 = I found the practice questions very difficult

Update the table as your revision progresses.

Revising at home

- Revise regularly — do *not* leave revision until near the examination.
- Plan your revision carefully so that there is no last-minute rush.
- Revise in a quiet room — you cannot revise properly if distracted by the television or music.
- Revise in short stretches — work for half an hour, have a break for 10 minutes, then start again. You should be able to revise for about 2–3 hours in an evening.
- Revise actively — read a topic, then close your book and make a summary from memory. Then go back and see what you've missed.
- Do as many questions as possible from sample and past papers.

In the exam room

- Think before you write.
- Don't waste time copying out the question.
- Make a plan for longer answers.
- Think in paragraphs.
- Don't rush.
- Don't panic — if you can't do a question, go on to the next one.
- Check your spelling of words that are similar to others.

Content Guidance

This section provides an overview of the key terms and concepts covered in **Module 4: Energy, Control and Continuity**. The major facts that you need to learn are outlined and the principles you need to understand are explained. Some A2 questions will test recall while others will test understanding. For example, you could be given data about a particular animal species that you have never heard of and asked to explain how it evolved.

The content of Module 4 falls into 12 main areas:

(1) Energy supply
The relationships between photosynthesis, respiration and ATP production.

(2) Photosynthesis
The chemical reactions of photosynthesis and the location of these in chloroplasts.

(3) Respiration
The chemical reactions of respiration and the location of some of these reactions in mitochondria.

(4) Survival and coordination
The transfer of information by the nervous system and hormones.

(5) Homeostasis
The principles involved in the regulation of body temperature, blood sugar and blood water potential; and the removal of metabolic waste.

(6) Nervous coordination
The functioning of the eye, the propagation of nerve impulses and some effects of drugs on the nervous system.

(7) Analysis and integration
The functioning of the brain and the autonomic nervous system.

(8) Muscles as effectors
Antagonistic muscle action, the structure of muscle fibres and the mechanism of muscle contraction.

(9) Inheritance
Meiosis and fertilisation, sex determination and the mechanism of inheritance.

(10) Variation
The ways in which genetic make-up and the environment interact to produce variation.

(11) Selection and evolution
Natural selection and the formation of new species.

(12) Classification
The principles used to divide organisms into groups.

Energy supply

All the energy used by every organism on Earth originates from the sun. Animals can only utilise thermal energy from the sun; for example, most reptiles bask in the sun during the morning to raise their body temperatures.

Plants can absorb sunlight and use the light energy to convert water and carbon dioxide into sugars. Sugars are then converted into more complex molecules such as starch, fats and proteins.

When animals eat plants, the complex molecules are digested into simple ones. Some of these simple molecules can be broken down to release energy by the process called respiration. Respiration is essentially the reverse of photosynthesis.

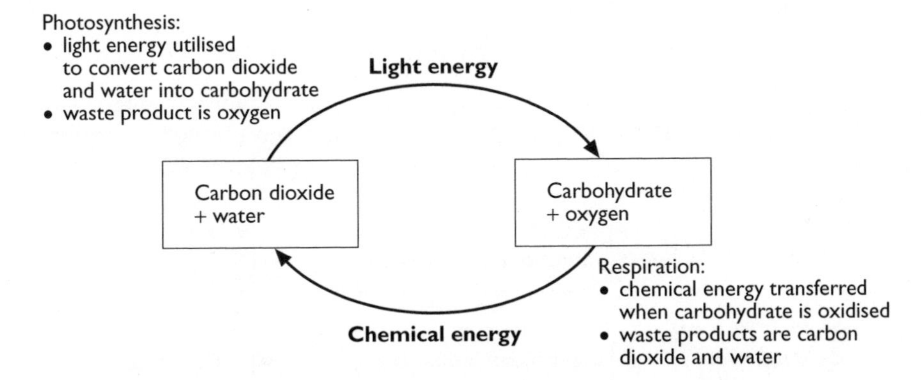

When substances burn, the energy in organic molecules is released in one go; most of this energy is released as heat. Living organisms use enzymes to break down sugars step by step, releasing energy in small, usable amounts. This energy is stored in **ATP**, ready for use. ATP releases energy for reactions that need it, such as muscular contraction.

ATP (adenosine triphosphate) contains three phosphate groups. The energy released from the respiration of sugars is used to add inorganic phosphate (P_i) to adenosine diphosphate (ADP) to produce ATP. ATP is also formed during photosynthesis, when light energy absorbed by chloroplast pigments is transferred to chemical energy.

Adenine – Ribose – P_i – P_i + P_i $\rightleftharpoons$ Adenine – Ribose – P_i – P_i – P_i

All living organisms use ATP in active transport and synthetic reactions. Animals also use ATP for muscle contraction.

Take care

- Remember that energy is not created or destroyed, only transferred from one form to another.
- Respiration does not create energy; it releases it from glucose.
- Plants respire 24 hours per day, not just when it is dark.

Photosynthesis

The word equation for photosynthesis is:

carbon dioxide + water + [light energy] ⟶ carbohydrate + oxygen

However, this is only a summary of a large number of reactions. These reactions are divided into two sets: the light-dependent reaction and the light-independent reaction.

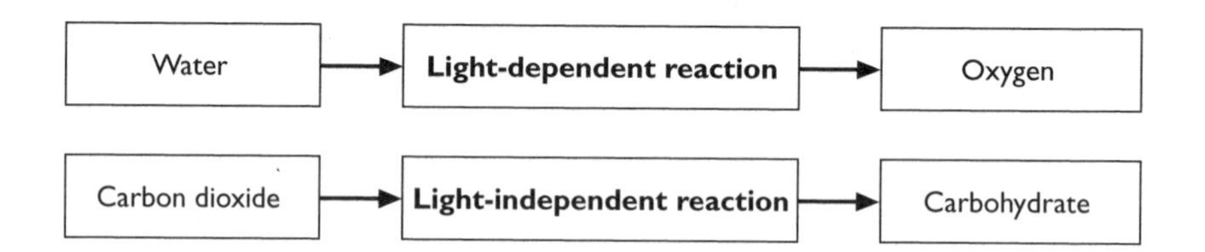

The essential feature of photosynthesis is that carbon dioxide is reduced to form carbohydrate. The **reduction** process involves electrons, so you need to understand the terms reduction and **oxidation** in terms of electrons.

Reduction	Oxidation
The gain of electrons	The loss of electrons
(The gain of hydrogen)	(The loss of hydrogen)
(The loss of oxygen)	(The gain of oxygen)

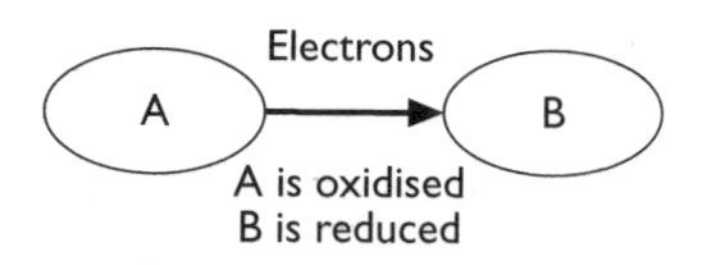

You also need to understand that electrons in chlorophyll molecules can be excited by light energy. When this happens they move into a higher electron shell.

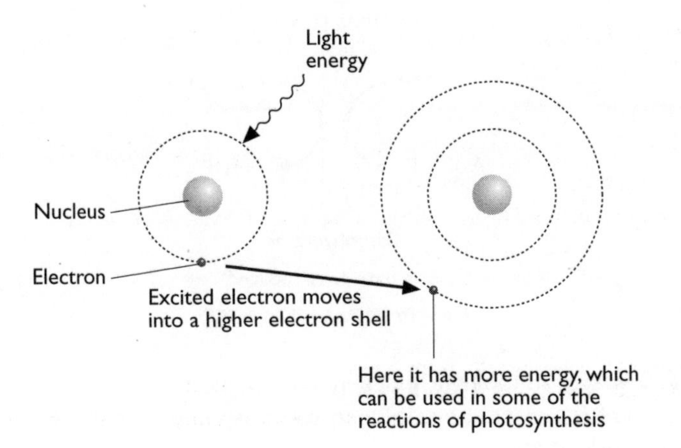

These 'high-energy electrons' can be used to reduce substances. They can also be used in other reactions.

Light-dependent reaction of photosynthesis

In the light-dependent reaction, energy from excited electrons is used:

- to generate ATP from ADP and P_i. This process is called **photophosphorylation**. Do not confuse this method of generating ATP with respiration
- to generate a very useful reducing agent called reduced **NADP** (nicotinamide adenine dinucleotide phosphate)
- to split water molecules into protons, electrons and oxygen. This process is called **photolysis**

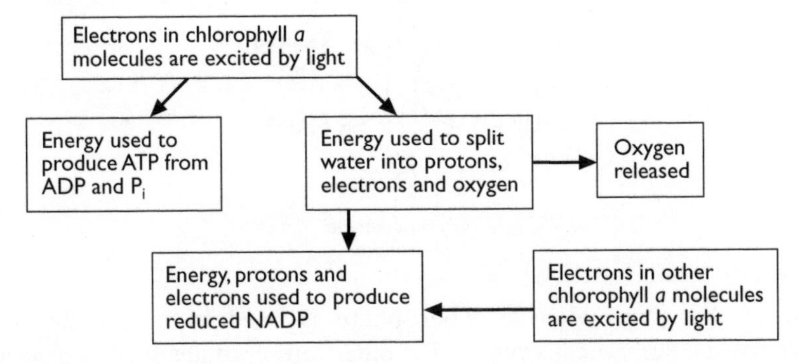

The end products of the light-dependent reaction are ATP and reduced NADP. Oxygen is released as a by-product.

Light-independent reaction of photosynthesis

In the light-independent reaction, the energy from ATP and the electrons from reduced NADP produced in the light-dependent reaction are used to reduce carbon dioxide from the atmosphere to form carbohydrate.

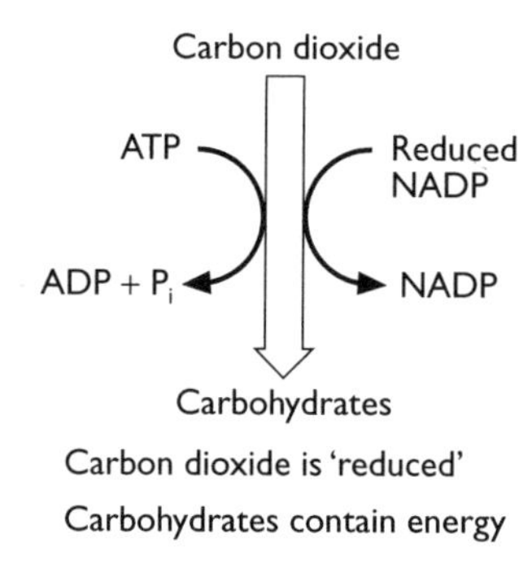

There are three stages in the light-independent reaction:

- carbon dioxide is accepted by **ribulose bisphosphate** to form two molecules of **glycerate 3-phosphate**

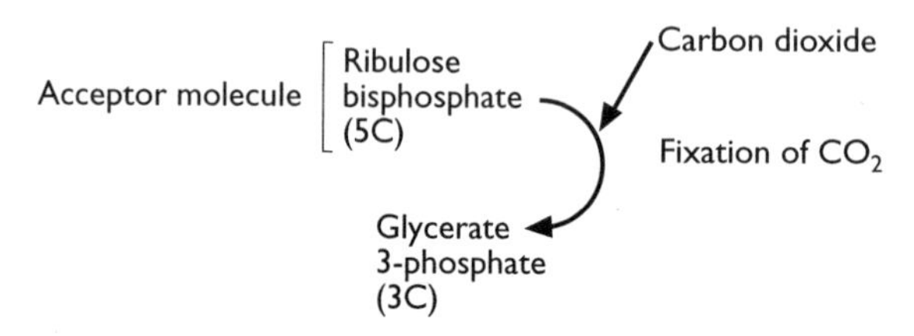

- the ATP and reduced NADP, produced by the light-dependent reaction, are used to reduce glycerate 3-phosphate to a carbohydrate called **glyceraldehyde 3-phosphate**

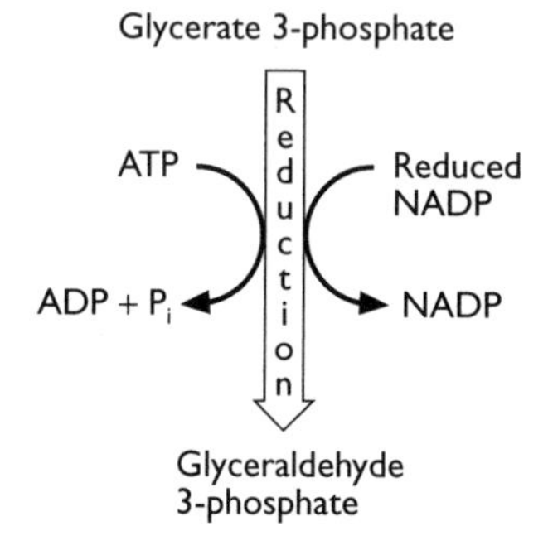

- most of the glyceraldehyde 3-phosphate is used to regenerate ribulose bisphosphate, but some is converted to other carbohydrates such as glucose and starch

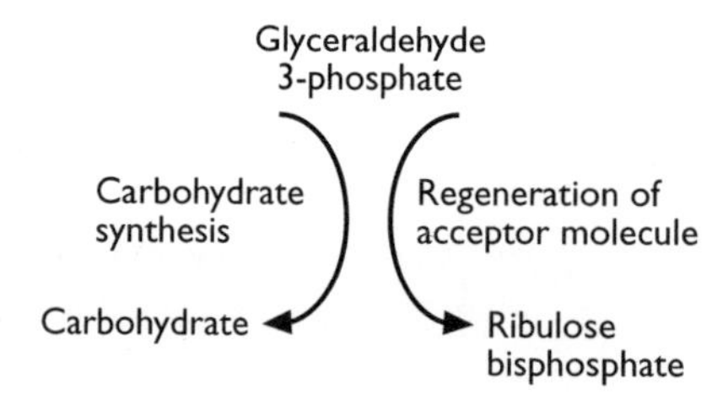

The role of chloroplasts in photosynthesis

Make sure that you revise the structure of the chloroplast from your AS notes.

The chloroplast is formed mainly from double phospholipid membranes called **thylakoids**. These membranes support the pigments and enzymes for the light-dependent reaction. Holding the pigments and enzymes close together in the membranes increases the efficiency of energy transfer. In places, thylakoid membranes are grouped in stacks called **grana**. The rest of the chloroplast is the **stroma**, where the light-independent reaction of photosynthesis takes place.

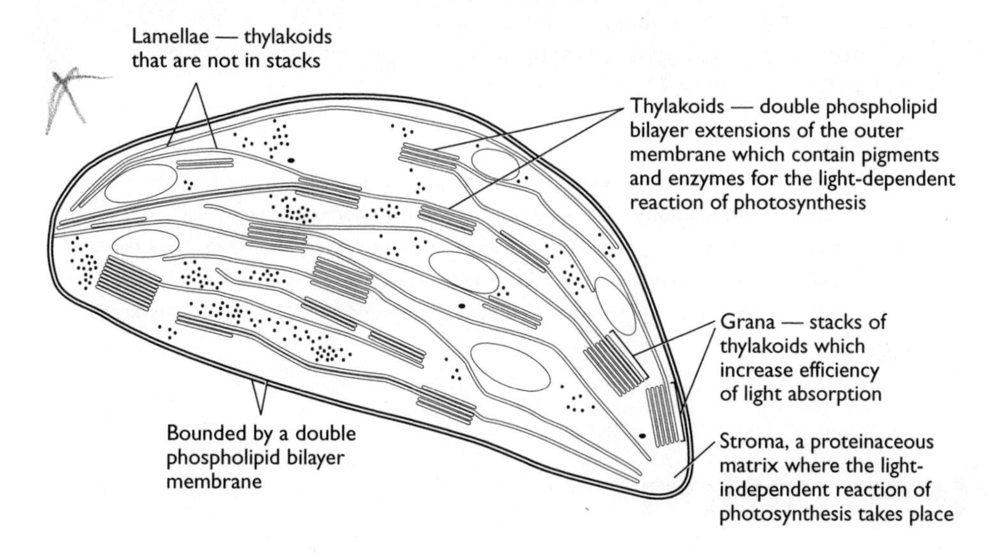

Linking the light-dependent and light-independent reaction

Be prepared for data linking the two sets of reactions.

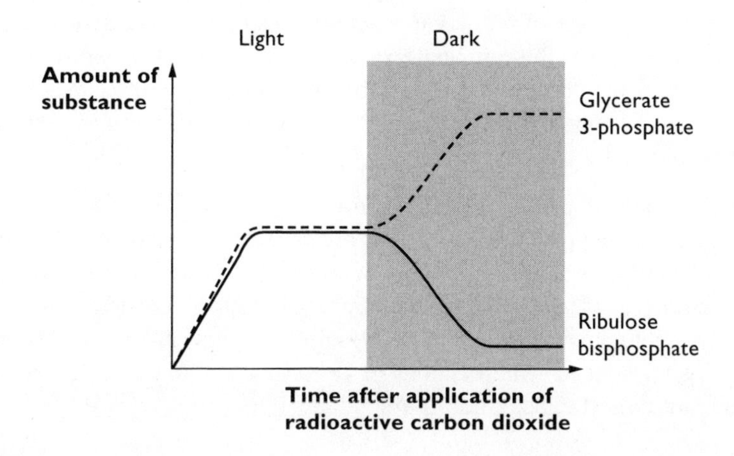

The graph on the previous page shows the following features:

- When the light is switched on, the amounts of ribulose bisphosphate and glycerate 3-phosphate rise as ATP and reduced NADP, produced by the light-dependent reaction, are used to produce glyceraldehyde 3-phosphate, some of which is used to produce more ribulose bisphosphate to act as an acceptor molecule for carbon dioxide fixation, producing more glycerate 3-phosphate.
- The amounts of glycerate 3-phosphate and ribulose bisphosphate become constant as the rest of the glyceraldehyde 3-phosphate is used to produce other carbohydrates.
- When the light is switched off, the supply of ATP and reduced NADP needed to reduce glycerate 3-phosphate disappears. Consequently, the amount of glycerate 3-phosphate rises and the amount of ribulose bisphosphate falls since there is no glyceraldehyde 3-phosphate to be recycled.

Take care

You need not learn the names of any intermediate compounds in photosynthesis other than the ones described above. You need not learn any chemical formulae. Many students try to learn detail that is not in the specification, with the result that they 'miss the wood for the trees'. Revise chromatography from your AS notes. You may be presented with chromatograms that show the sequence of the substances formed in the light-independent reaction.

Respiration

The word equation for aerobic respiration is:

$$\text{glucose} + \text{oxygen} \longrightarrow \text{carbon dioxide} + \text{water} + [\text{energy}]$$

However, this is only a summary of a large number of reactions. Whereas photosynthesis is essentially a set of reduction reactions, respiration is a series of oxidising reactions. In respiration, the principal oxidising agent is **NAD**. When this oxidises a compound it becomes reduced NAD. (Not to be confused with reduced NADP. To remember which is which, remember the P — NADP is used in photosynthesis, NAD is used in respiration.)

There are four main stages in aerobic respiration:

- **glycolysis**, in which glucose molecules are split into two molecules of **pyruvate**, each containing three carbon atoms. To begin this series of reactions, two ATP molecules are used. But later reactions produce four molecules of ATP. There is therefore a net gain of two ATP molecules for each molecule of glucose. Reduced NAD is also produced. The reactions of glycolysis take place in the endoplasmic reticulum of the cytoplasm, not in the mitochondria.

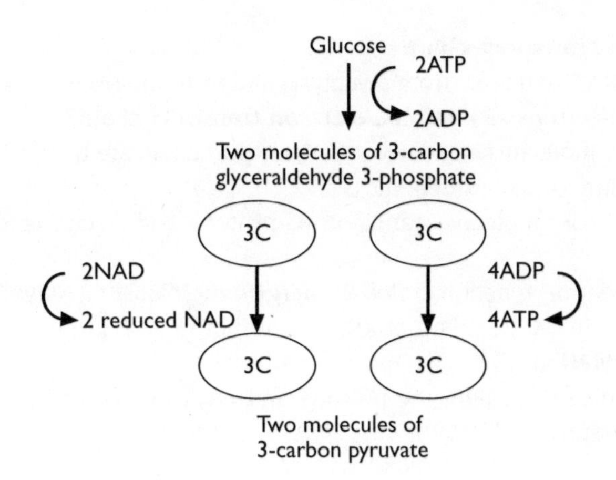

- **the link reaction**, in which pyruvate combines with **coenzyme A** to produce **acetylcoenzyme A**. In this reaction, the pyruvate is oxidised, and also loses a molecule of carbon dioxide.

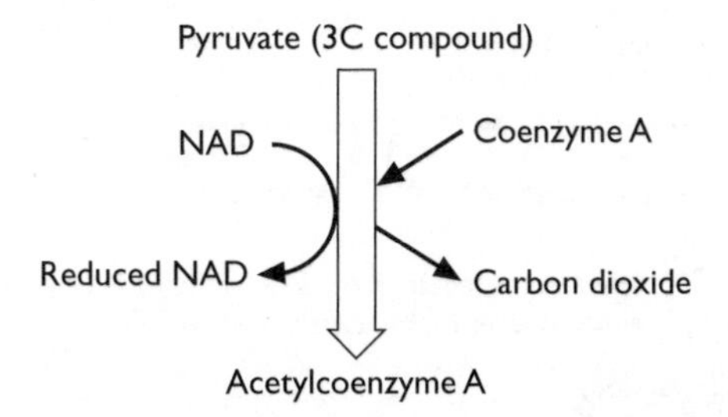

- **the Krebs cycle**, which is a series of oxidation reactions resulting in the production of reduced NAD and a similar compound, reduced **FAD**. Acetylcoenzyme A contains two carbon atoms from the original glycerate. It combines with a 4-carbon compound, producing a 6-carbon compound. As the6-carbon compound is oxidised, first to a 5-carbon compound and then to re-form the 4-carbon compound, two molecules of carbon dioxide are released.

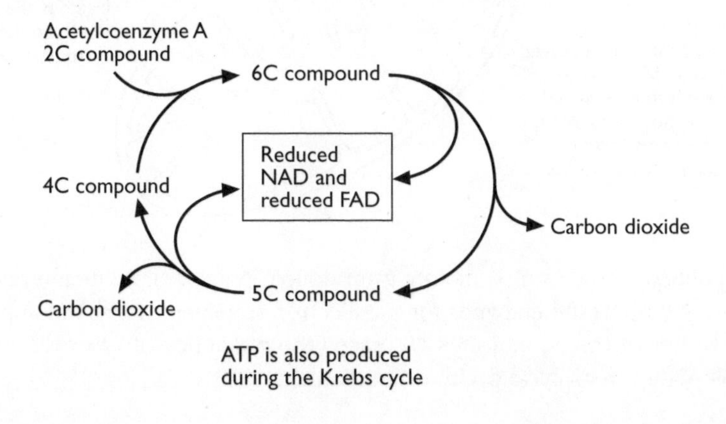

- **the electron transport chain**
  (1) Reduced NAD and FAD (from glycolysis and from the Krebs cycle) feed protons (H^+) and electrons (e^-) into the **electron transport chain**.
  (2) The carrier molecules in the electron transport chain are held in close proximity in the cristae of the mitochondria.
  (3) As each carrier molecule hands on its protons and electrons to the next it is oxidised.
  (4) These oxidation reactions release energy which is eventually used to convert ADP + P_i into ATP. This method of producing ATP is known as **oxidative phosphorylation**.
  (5) At the end of the chain, the protons and electrons combine with oxygen to produce water.

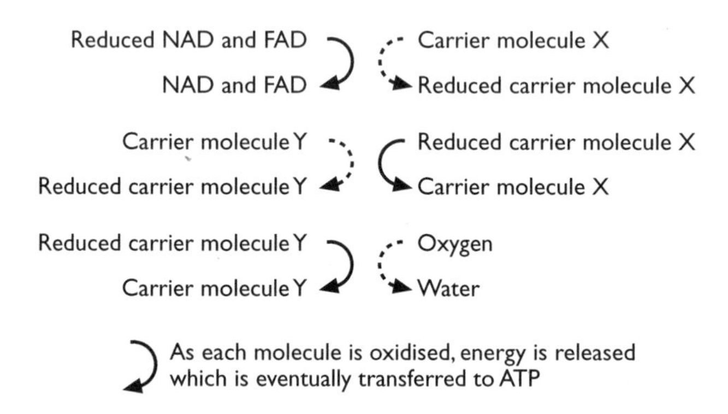

The role of mitochondria in respiration

Make sure that you revise the structure of the mitochondrion from your AS notes.

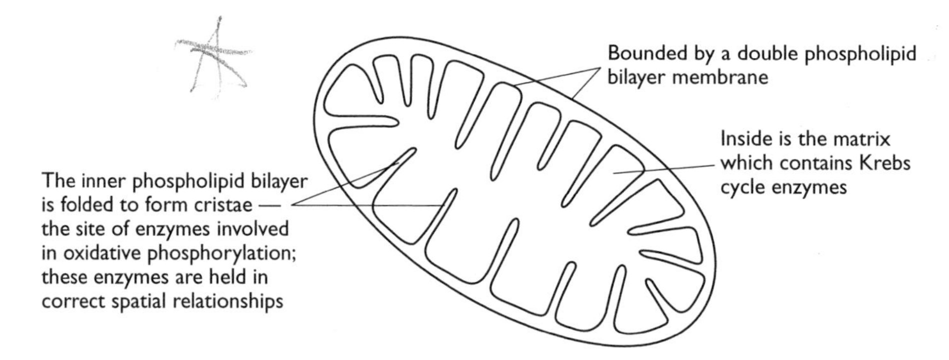

The mitochondrion is formed mainly from double phospholipid membranes. These membranes support the enzymes for the electron transport chain. Holding enzymes close together in the membranes increases the efficiency of energy transfer. The reactions of the Krebs cycle occur in the matrix.

Comparing ATP production in photosynthesis and respiration

Photophosphorylation (photosynthesis)	Oxidative phosphorylation (respiration)
In thylakoids of chloroplasts	In cristae of mitochondria
Energy source is electrons excited by light	Energy source is transfer of electrons during oxidation reactions
NADP is the electron acceptor	NAD is the electron acceptor
Chlorophyll is necessary	No chlorophyll in mitochondria
Most ATP produced is used in the light-independent reaction	ATP is used in a wide variety of reactions

Take care

Do not try to learn the names or the formulae of substances that are not listed above. You need to understand the processes rather than learn the names of lots of chemicals.

Survival and coordination

Organisms can respond to changes in their environment. This might increase their chances of survival.

The diagram on page 22 shows the mechanism of one type of response to a **stimulus**. The stimulus is the hot plate and the **response** is to jerk the arm away from the source of heat. In this example, the response is coordinated by the nervous system. The structures involved make up a **reflex arc**. This reflex arc has three neurones:

- a **sensory neurone**, which transmits impulses from the receptor to the spinal cord
- a **motor neurone**, which transmits impulses from the spinal cord to the **effector**
- a **relay neurone**, which carries impulses from the sensory neurone to the motor neurone

Information is passed from one neurone to the next via minute gaps called **synapses**.

Stimuli are detected by **receptors**. There are many different types of receptor, each specific to a particular type of stimulus. For example, the retina of the eye contains receptors sensitive to light; the skin contains receptors sensitive to touch and other receptors sensitive to temperature changes.

The structures that bring about the response are called effectors. Effectors can be muscles, as in the pain withdrawal reflex illustrated below, or glands. One example of glands acting as effectors is the salivary glands, which make your mouth 'water' when you smell delicious food.

The response is often coordinated by the nervous system, but some responses can be coordinated entirely by glands.

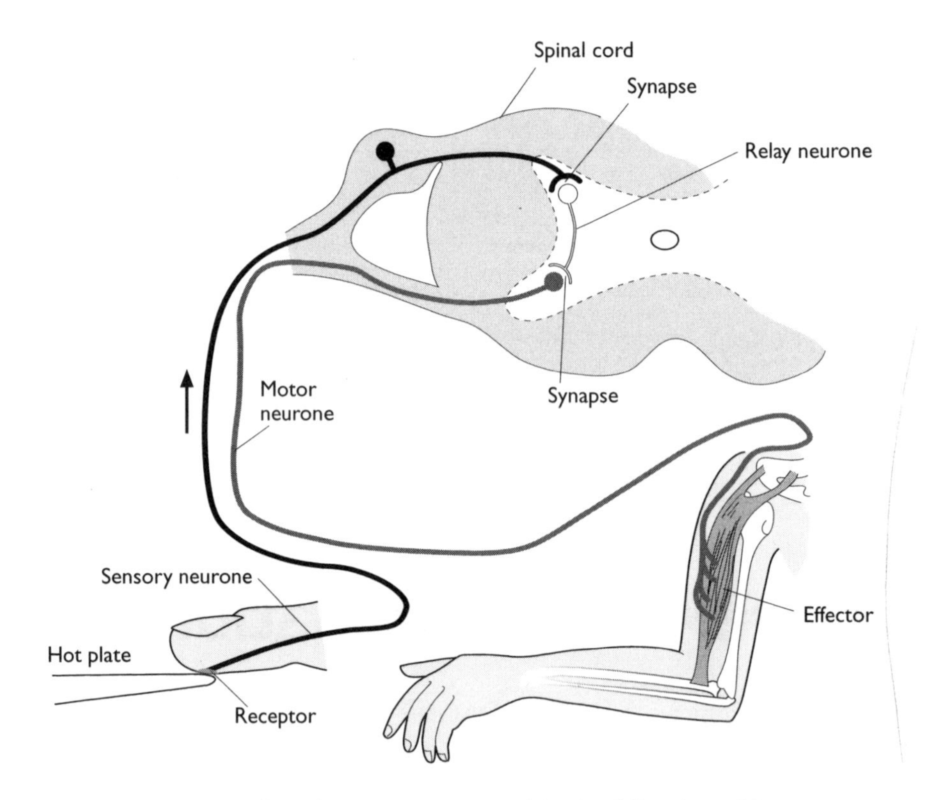

The pain withdrawal reflex above is summarised in the following table:

Stage	**Example**
Stimulus	Hot plate
Receptor	Temperature receptor
Coordinators	Sensory neurone Synapse Relay neurone Synapse Motor neurone
Effector	Muscle
Response	Muscle contracts and moves arm

You might be given any example of a reflex action to analyse in terms of stimulus, receptor, coordinator, effector and response. Try practising on the following:

- knee jerk reflex
- pupil reflex

- mouth watering reflex
- blinking reflex

Take care
Make sure that you do not give 'brain' as the coordinator for all examples of reflex arcs. If both receptor and effector are below the head, the coordinator is probably the spinal cord.

Homeostasis

Homeostasis involves keeping conditions inside an organism constant. Some of the reasons why this is important are:

- changes in temperature and pH affect activity of enzymes — extremes can lead to denaturation
- a constant internal temperature enables mammals to be independent of external fluctuations in temperature
- maintaining constant water potential avoids osmotic problems

Negative feedback

Many conditions in the human body have set resting levels, for example:

- body temperature 36.5–37.2°C
- plasma pH 7.4–7.5
- plasma sodium 135–143 mmol $litre^{-1}$
- plasma glucose (fasting level) 3.5–7.5 mmol $litre^{-1}$

Fluctuations from these resting levels are detected by relevant receptors. Effectors then bring the levels back within the resting range. The process that returns conditions to the resting level is called **negative feedback**. This is summarised below.

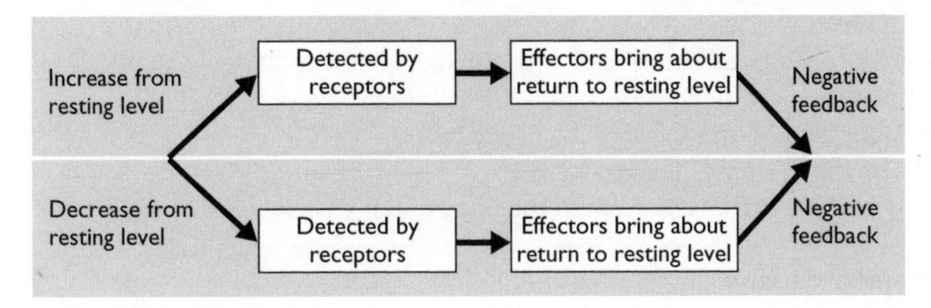

Temperature regulation

Mammals and birds maintain a relatively constant, high body temperature using a variety of mechanisms. They are known as **endothermic** animals, since control comes from within the body. Other vertebrates are **ectothermic** — they rely on the external environment to maintain a suitable temperature. For example, crocodiles bask in the sun to warm up and return to water to cool down.

The diagram below summarises the processes that occur to control body temperature in mammals. Note particularly that there are different receptors for detecting increase and decrease in temperature.

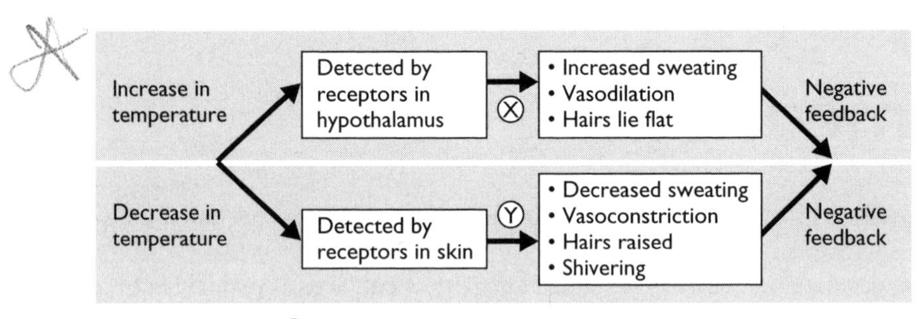

Ⓧ — coordinated by heat loss centre in hypothalamus
Ⓨ — relayed to heat conservation centre in hypothalamus

Sweating

- The **thermoregulatory centre** in the brain is sensitive to a rise in the temperature of blood.
- The effectors are the sweat glands.
- The rate of sweating increases.
- Evaporation of sweat cools the body.

Take care

Sweating does not cool the body; cooling occurs by the *evaporation* of sweat.

Vasodilation

- The thermoregulatory centre in the brain is sensitive to a rise in the temperature of blood.
- The effectors are arterioles — those supplying capillaries near the skin surface dilate.
- More blood flows through the capillaries.
- The skin becomes flushed (red).
- More heat is transferred from the body to the environment by radiation.

Vasoconstriction

- Receptors in the skin detect a decrease in temperature.
- The effectors are arterioles — those supplying capillaries near the skin surface constrict.
- Less blood flows through capillaries.
- The skin becomes pale.
- Less heat is transferred from the body to the environment.

Take care

Blood vessels do *not* move up and down in the skin; capillaries do *not* constrict or dilate.

Shivering

- Receptors in the skin detect a decrease in temperature.
- The effectors are skeletal muscles — these muscles contract involuntarily.
- This results in an increased rate of respiration in muscle fibres.
- Thermal energy is transferred to blood flowing through muscles.

Hair erection

- Receptors in the skin detect a decrease in temperature.
- The effectors are tiny muscles attached to the base of the hair follicles — these muscles contract, pulling the hairs upright.
- The hairs trap an insulating layer of air.
- This mechanism has a negligible effect on temperature in humans, but is very effective in mammals with fur.

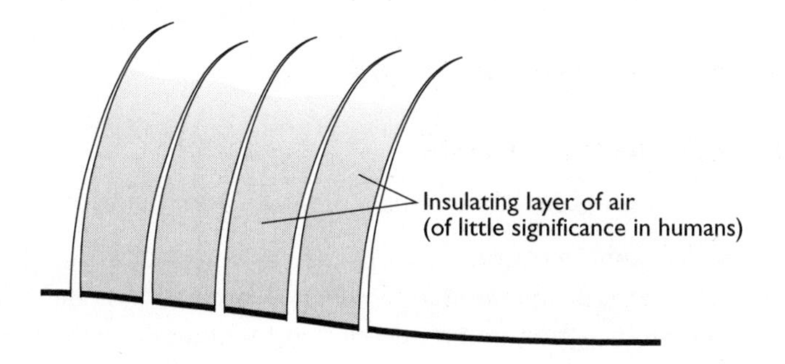

Blood glucose regulation

Blood glucose concentration is controlled by the pancreas.

You studied the secretion of digestive enzymes in AS unit 1. Besides secreting enzymes, the pancreas has groups of cells called **islets of Langerhans**. These cells secrete hormones that control blood glucose concentration. **β-cells** in the islets detect increases in blood glucose concentration. They respond by secreting the hormone **insulin**. Insulin stimulates glucose carrier proteins in the cytoplasm of body cells to move to the cell surface membrane. The increased number of carrier molecules leads to an increase in the quantity of glucose that can be absorbed. Insulin also activates enzymes in the liver to convert glucose into glycogen. This decreases the blood glucose concentration of blood passing through the liver.

α-cells in the islets detect decreases in blood glucose concentration. They respond by secreting the hormone **glucagon**. Glucagon activates enzymes in the liver to convert glycogen back into glucose. This increases the blood glucose concentration of blood passing through the liver. These processes are summarised in the diagram below.

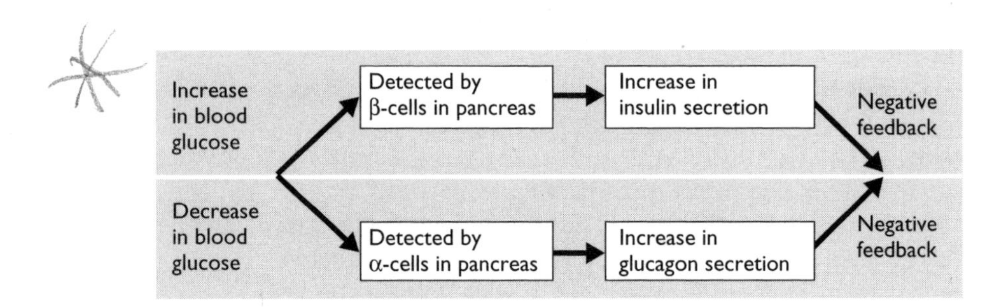

Take care

- The brain is *not* involved in blood sugar regulation.
- The pancreas secretes *hormones* to control blood glucose, *not* enzymes.
- Insulin does *not* convert glucose to glycogen.
- Glucagon does *not* convert glycogen to glucose.
- However, insulin and glucagon activate the enzymes that convert glucose to glycogen and glycogen to glucose respectively.
- The liver secretes enzymes that convert glucose to glycogen and glycogen to glucose, *not* hormones.
- Do not confuse the spelling of glycogen and glucagon.

Removal of metabolic waste

Excretion is the removal of metabolic waste products from the body.

Excess amino acids cannot be stored (as opposed to excess glucose which is stored as glycogen), so they are deaminated in the liver. **Deamination** is the removal of an amino group. The amino group is first converted to ammonia which is very toxic. Ammonia is therefore converted to a less toxic compound called urea. Urea is transported via the blood to the kidneys where it is removed from the blood during the formation of urine.

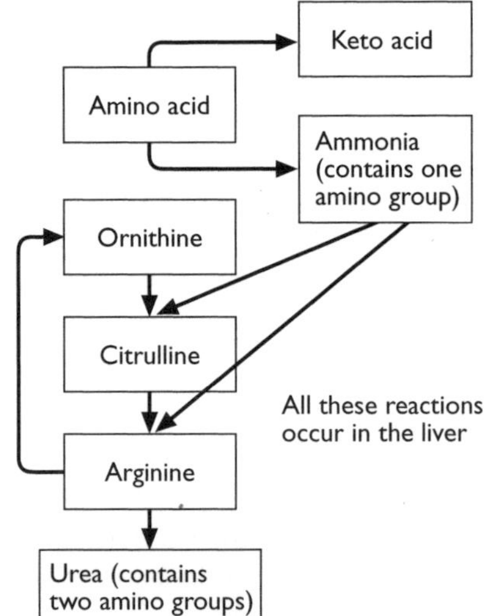

Urine is produced by the kidneys. There are two essential processes in urine formation:

- **ultrafiltration**
- **selective reabsorption**

Ultrafiltration

- Ultrafiltration is caused by high hydrostatic pressure of blood forcing part of the plasma out of the capillaries in the **glomeruli**.
- The filtrate has the same composition as tissue fluid, i.e. blood plasma minus proteins (which are too large to pass through capillary walls).

Selective reabsorption

- This occurs via intrinsic protein molecules in the outer membranes of cells in the first convoluted region of the kidney.
- These specific protein molecules have specific receptor site shapes for each type of solute, which accounts for selective permeability.
- Energy from respiration is required.
- This gives potential for movement against a concentration gradient.
- *All* glucose and most of the ions in the filtrate are reabsorbed into the blood by active transport.
- Most of the water in the filtrate is reabsorbed via osmosis.
- *Some* urea is also reabsorbed.

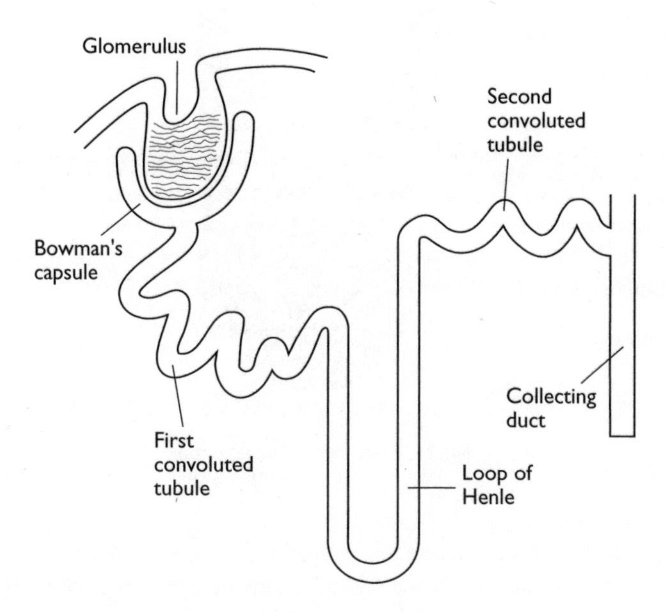

Regulation of blood water potential

The rest of the kidney tubule is concerned mainly with the regulation of blood water potential. The first stage in this process is the production of an ion gradient across the medulla of the kidney by the **loop of Henle**. This process is shown in the two diagrams below.

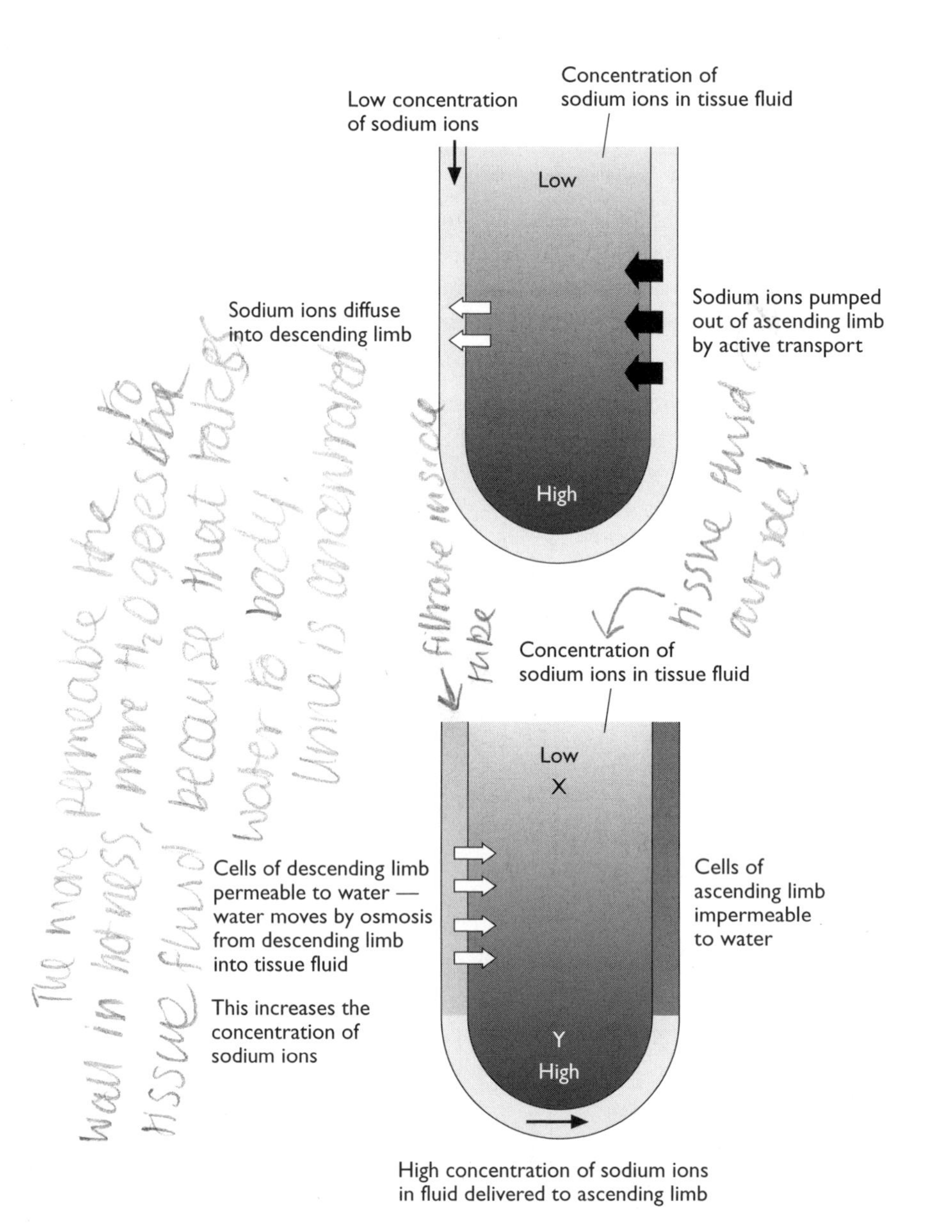

As the process is repeated, the concentration gradient of sodium ions across the medulla, from X to Y, becomes steeper. The maintenance of this gradient is important for the reabsorption of water from the remaining filtrate in the second convoluted region and collecting ducts of the kidney tubule.

How much water is reabsorbed depends on the amount of anti-diuretic hormone (**ADH**) in the blood. ADH is secreted when the water potential of the blood becomes more negative. This happens when we lose a lot of water via sweating, for example. However, if the water potential of the blood becomes less negative, as when we drink a lot of water, then ADH is not secreted. The figure below shows the sequence of events leading to the production of ADH.

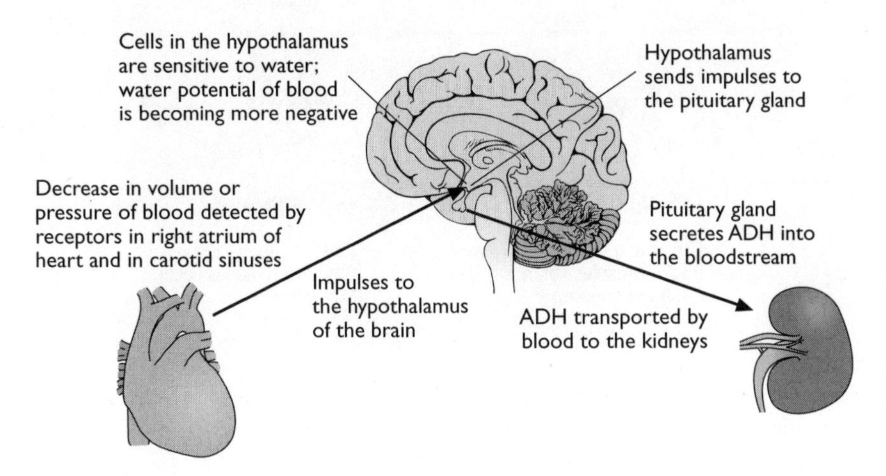

ADH affects the permeability of the walls of the second convoluted region and the collecting ducts. If these walls become more permeable, then water will leave the filtrate by osmosis. This is because of the high concentration of ions in the medulla of the kidney, produced by the loop of Henle. The urine produced in these circumstances is scant and concentrated. If ADH is not secreted, the walls become impermeable, water does not leave the filtrate and the urine is copious and dilute.

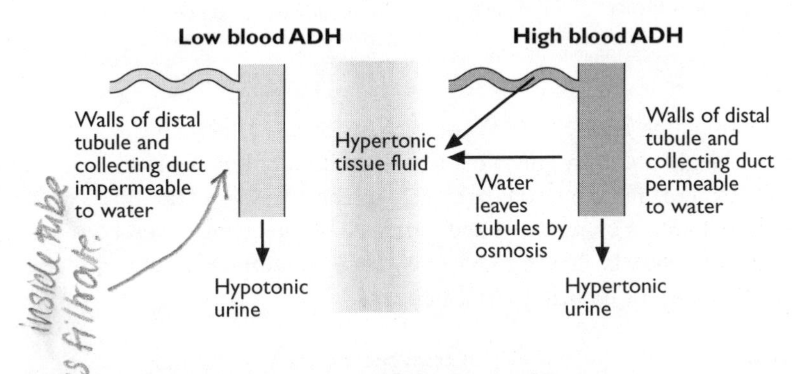

Nervous coordination

The mammalian eye

Make sure that you can label the parts of the eye from your GCSE notes.

The role of the iris

The amount of light that enters the eye is controlled by the **iris**. The iris contains two sets of antagonistic involuntary muscles — circular and radial. These constrict and dilate the pupil respectively.

The adjustment of the size of the pupil is a reflex action.

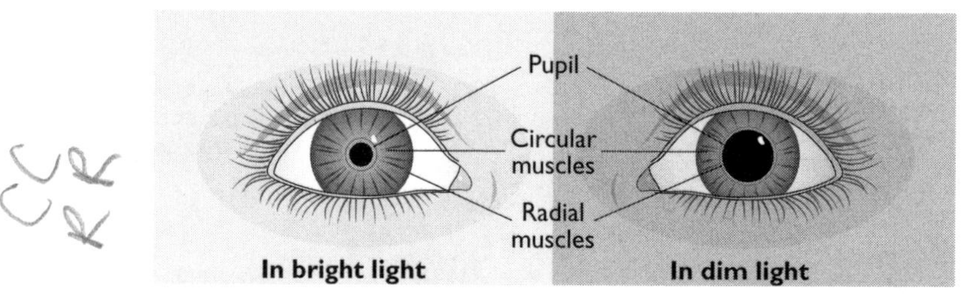

Circular muscles contract
Radial muscles relax
Pupil constricted

Radial muscles contract
Circular muscles relax
Pupil dilated

Focusing

Focusing occurs via refraction — the change in the speed of light as it passes from one medium into another. Most **refraction** occurs at the interface between air and the cornea. The lens refracts the light only slightly.

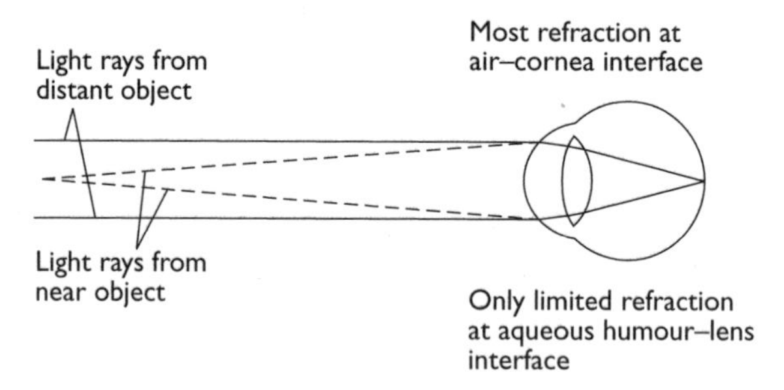

Accommodation is the ability to focus objects which are at different distances from the eye. It is brought about by contraction or relaxation of the **ciliary** muscles. When the eye is at rest, the ciliary muscles are relaxed. The pressure of the fluids in the eye forces the eyeball into a spherical shape. The wall of the eyeball exerts a tension on the **suspensory ligaments**. These in turn pull the lens into a less convex shape with a longer focal length for focusing distant objects.

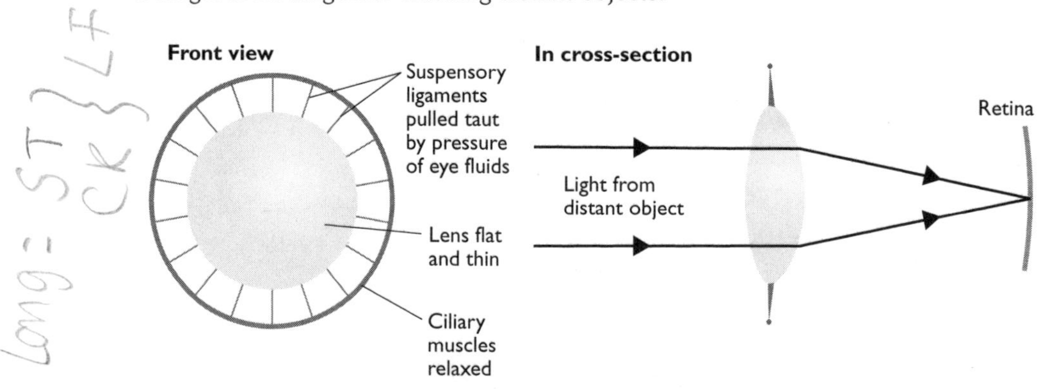

In most parts of the body, contraction of a muscle increases tension in the structure to which the muscle is attached. But, because the ciliary muscles form a ring, contraction of the ciliary muscles pulls the wall of the eyeball inwards and this results in less tension in the suspensory ligaments. The lens now becomes more convex with a shorter focal length for focusing near objects.

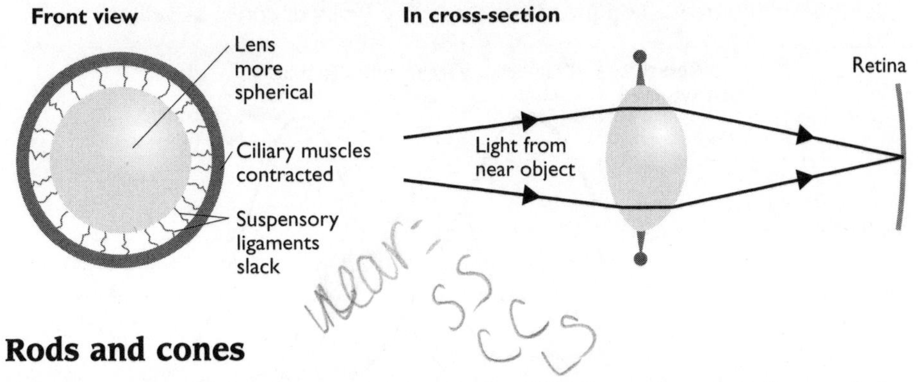

Rods and cones

Rods and **cones** are the light-receptor cells found in the **retina**. The diagram below shows the structure of a rod cell. A cone cell has the same basic structure, but is rounder and contains the pigment **iodopsin** rather than **rhodopsin**.

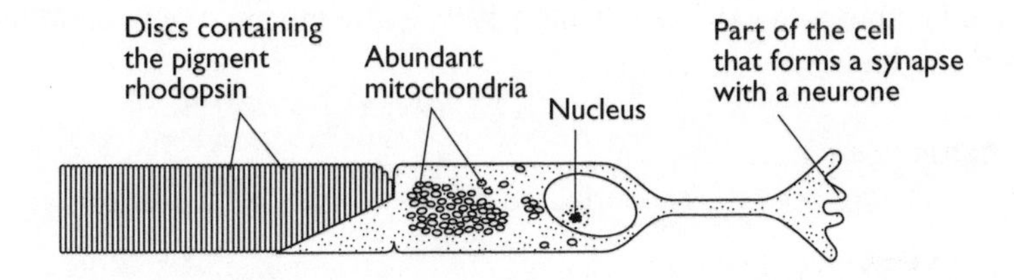

Each rod contains the light-sensitive pigment rhodopsin arranged in discs. In light, the rhodopsin molecule breaks down into a pigment called retinal and a protein called opsin. The breakdown of rhodopsin into opsin results in the release of a transmitter substance by the rod cell. This transmitter substance passes across the synapse between the rod cell and a neurone.

In the dark, rhodopsin is resynthesised from retinal and opsin. This is called dark adaptation and explains why we are able to see more as we get used to the dark. In the dark, it takes about 30 minutes to resynthesise all the rhodopsin. The resynthesis of rhodopsin requires ATP which is supplied by the abundant mitochondria.

Rod cells are sensitive to very low light intensities, while cone cells are only sensitive to high light intensities.

Colour vision

Cone cells enable us to see colours. There are three types of cone: red-light receptors, green-light receptors and blue-light receptors. The ranges of sensitivity of these three types overlap and most wavelengths of light stimulate at least two types of cone. One theory for the mechanism of colour vision is the **trichromatic** theory. This theory states that we see colours by a mixing of the three primary colours — blue, green and red. A white object reflects all the colours of the spectrum; this stimulates all three types of cone and the brain interprets the impulses from the cones as white light. Yellow light will stimulate both red-light cones and green-light cones, so the brain interprets simultaneous impulses from these two types of cones as yellow.

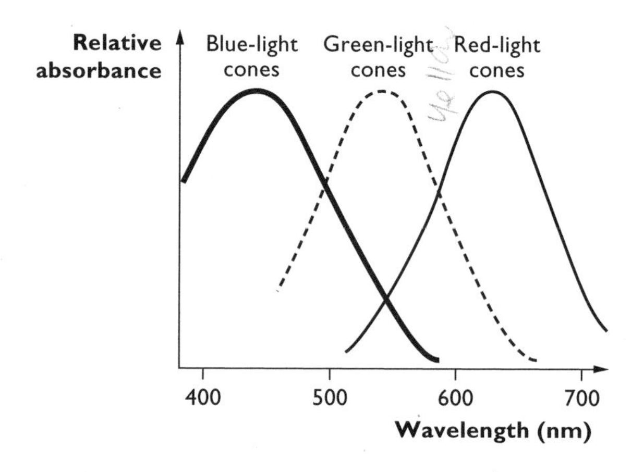

You should be able to recall the colours produced by mixing any two of the colours red, green and blue.

Retinal convergence

Rods and cones are not evenly distributed across the retina (see below).

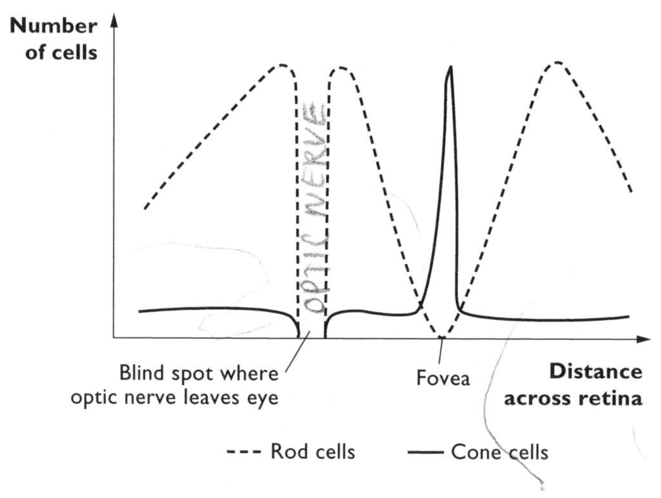

- The **fovea**, where the image is focused in bright light, consists almost entirely of cones.
- The rest of the retina consists mainly of rods.

- There are neither rods nor cones at the blind spot, where the optic nerve leaves the eye.

The diagram below shows how rods and cones are connected to the sensory neurones in the optic nerve.

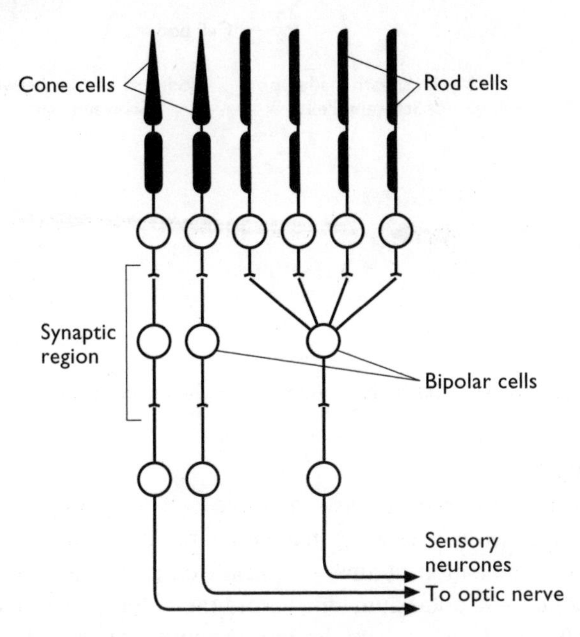

At the fovea, the majority of **bipolar cells** synapse with a single cone cell. This results in a high visual acuity (sharpness) for images focused here. Images are only focused here in bright light.

In the rest of the retina, each bipolar cell synapses with several rod cells. This means that in low light intensity, several rod cells together will produce sufficient transmitter substance to cause the bipolar cell to send an impulse to the brain. This results in high sensitivity for the rod-containing parts of the retina.

The nerve impulse

The structure of myelinated nerve fibres

Myelinated neurones consist of:

- a cell body which contains the cell nucleus
- **dendrites** which synapse with other neurones
- an **axon** which carries impulses away from the cell body
- a **myelin sheath**. This is built up of **Schwann cells** which contain large amounts of fatty material. The junctions between Schwann cells are called **nodes of Ranvier**. The myelin sheath allows an impulse to 'jump' from node to node, giving a far higher conduction speed than in **non-myelinated** neurones.

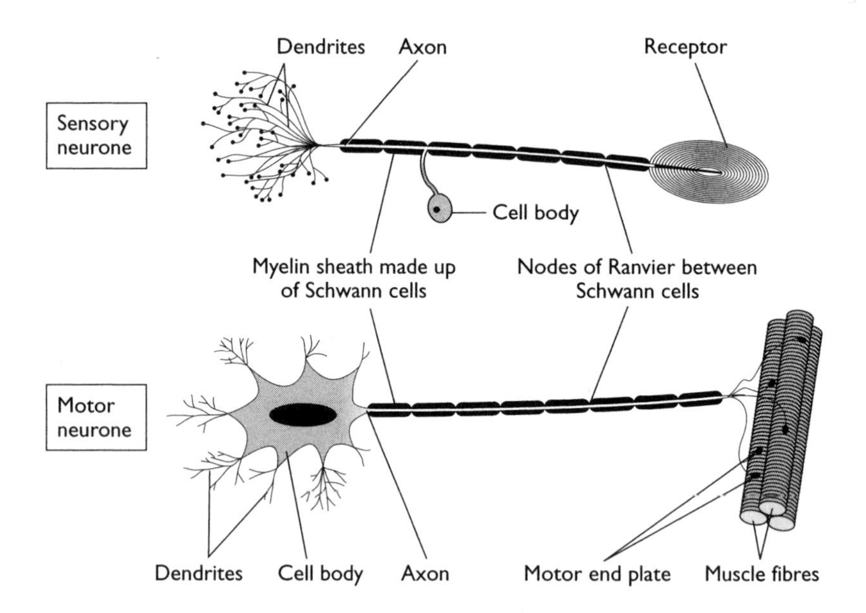

Resting potential

In order to conduct a nerve impulse, a nerve cell must first develop a **resting potential**. There are two mechanisms involved in producing the resting potential:

- active transport by cation pumps — specialised carrier protein molecules that transport sodium and potassium ions across the outer membrane of the neurone
- facilitated diffusion which occurs through channel protein molecules

The cation pump includes the enzyme ATPase, which is also a carrier protein.

The two ions also move through the axon membrane by facilitated diffusion through channel proteins. However, the channel proteins move ions in the opposite direction to the cation pump:

- Na^+ ions diffuse in through the membrane and K^+ ions diffuse out
- the channel proteins that carry K^+ ions outnumber those that carry Na^+ ions, so more K^+ ions diffuse out than Na^+ ions diffuse in
- the cation pumps move more ions than are moved by facilitated diffusion

The net result is therefore that there are more sodium ions outside the membrane than potassium ions inside. This gives a net positive charge outside the membrane. This potential difference is known as the resting potential.

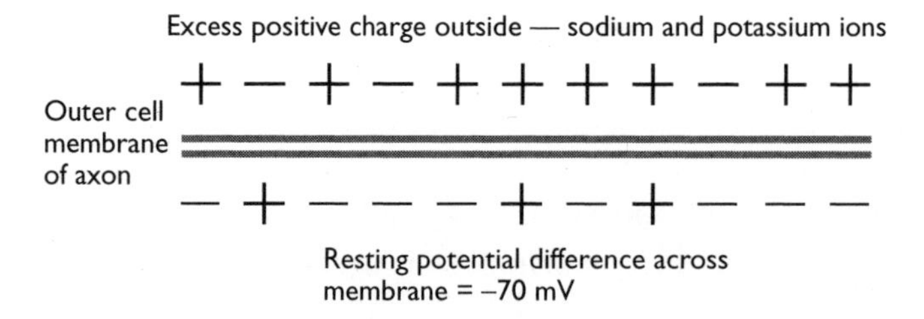

Action potential

When a neurone is stimulated, an impulse passes down the length of its axon. This impulse is a wave of **depolarisation**. One small stretch of axon after another develops a positive charge with respect to the fluid outside it.

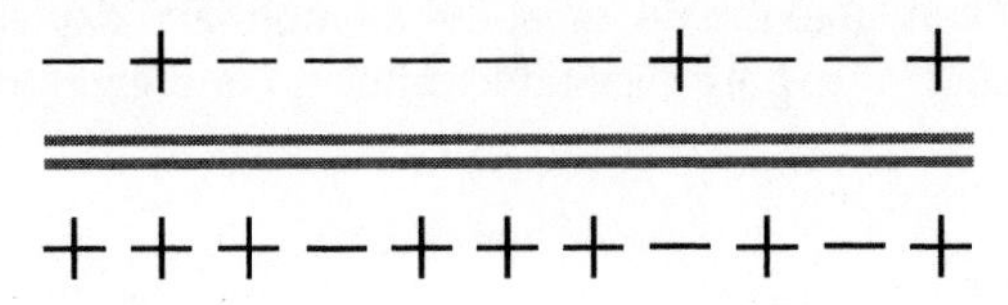

Depolarisation happens because of channels in the membrane of the axon that can change shape. In the 'open' position, they allow ions to pass through; in the 'closed' position, they block ion movement. Because they can be open or closed, these protein channels are called gated ion channels.

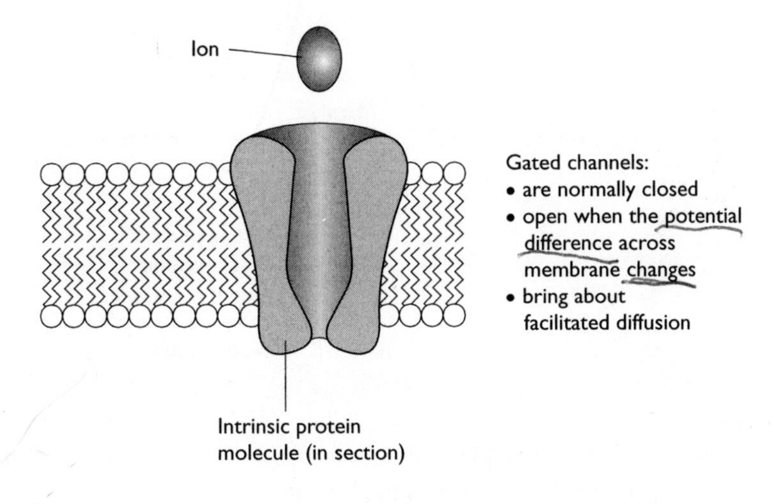

In depolarisation:

- there is a change in the polarisation of the membrane
- gated channels for sodium open
- sodium ions diffuse in along the concentration gradient by facilitated diffusion
- the inside of the membrane is now positively charged

In repolarisation:

- the gated channels for sodium close
- more gated channels for potassium open
- more potassium ions diffuse out
- the resting potential is quickly restored

Refractory period

This is a short period following depolarisation when the membrane cannot be depolarised. This separates nerve impulses from one another. It occurs when the gated sodium channels are closed during repolarisation.

The graph

Make sure that you can label the regions on the graph, and explain what happens at each in terms of active transport, facilitated diffusion and gated channels.

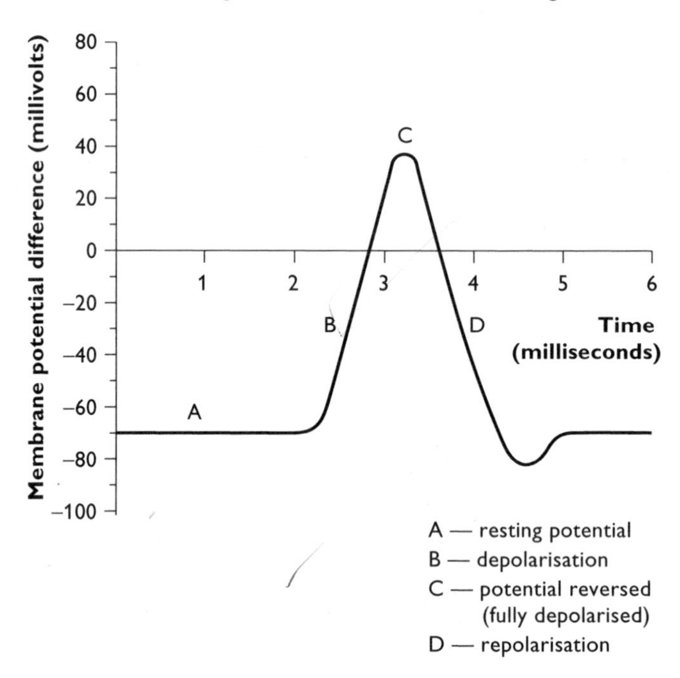

Summary

- An action potential is a wave of depolarisation passing along the axon of a neurone.
- It is started by a stimulus which reaches 'threshold'.
- It is 'all-or-nothing', i.e. it always has the same potential difference.
- A larger stimulus increases the *number* of action potentials in a given time but does *not* increase the size of the action potentials.

Synapses

Impulses are transmitted across synapses by **transmitter molecules**. The transmitter is **acetylcholine** at **cholinergic synapses** and **noradrenaline** at **adrenergic synapses**. The transmitter is stored in vesicles.

When released, the transmitter diffuses across the synapse and attaches to a receptor site on the post-synaptic cell, which then depolarises. The diagram below shows the sequence of events at a cholinergic synapse.

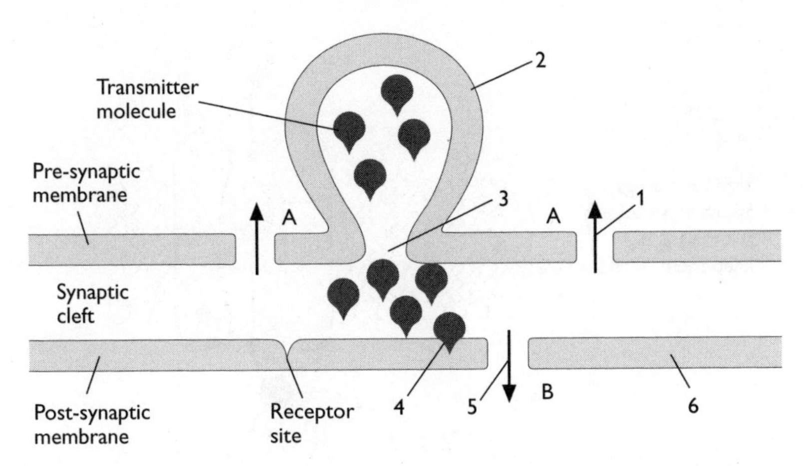

(1) Gated channels for calcium ions open
(2) Vesicles containing transmitter substance fuse to the **pre-synaptic membrane**
(3) Vesicles release acetylcholine
(4) Acetylcholine attaches to receptor sites in the **post-synaptic membrane**
(5) Gated channels for sodium ions open
(6) The post-synaptic membrane is depolarised

The diagram below shows a neuromuscular junction. Exactly the same sequence of events occurs when an impulse arrives at the pre-synaptic membrane, resulting in the depolarising of the muscle fibre outer membrane.

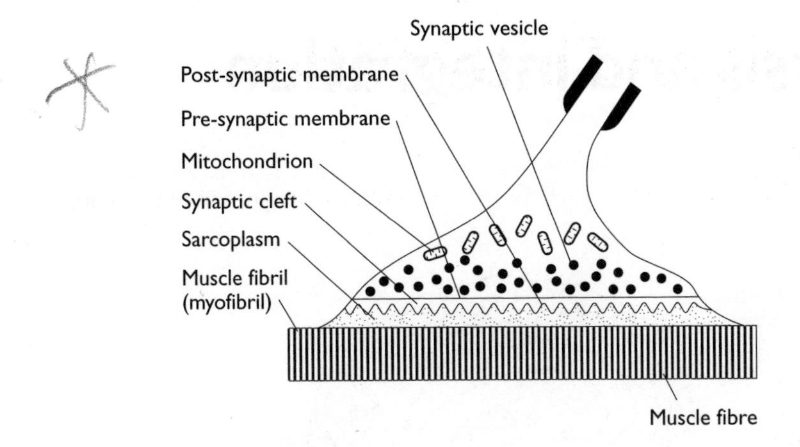

Drugs and synapses

Many drugs affect transmission across synapses. Most of them operate at gated channels. Stimulant drugs keep gated channels open so that the post-synaptic membrane is continually being depolarised. Depressant drugs keep gated channels closed, preventing depolarisation of the post-synaptic membrane, as shown in the diagram below.

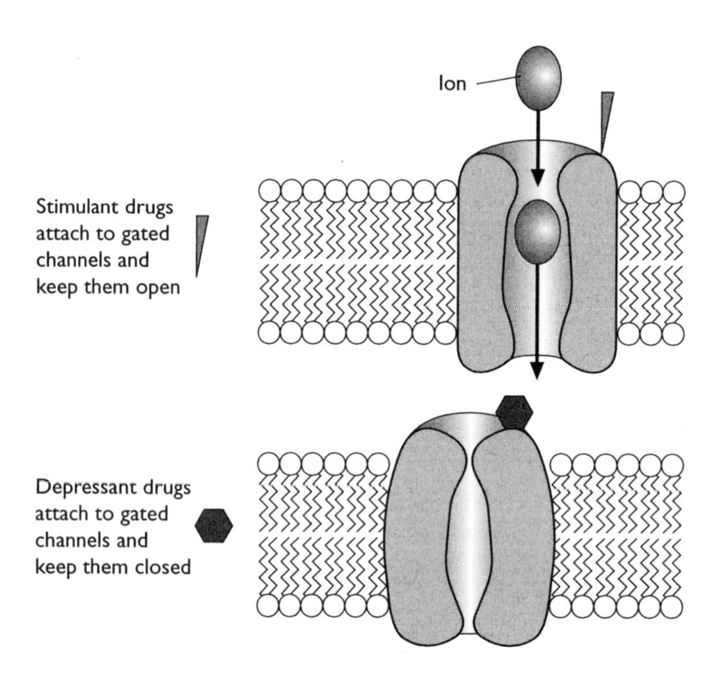

Be prepared for questions on drugs that you have not studied; the answer required will normally be in terms of the shapes of drug molecules and receptor sites on gated channel molecules.

Analysis and integration

The brain

The diagram below shows the parts of the brain referred to in different sections of this module.

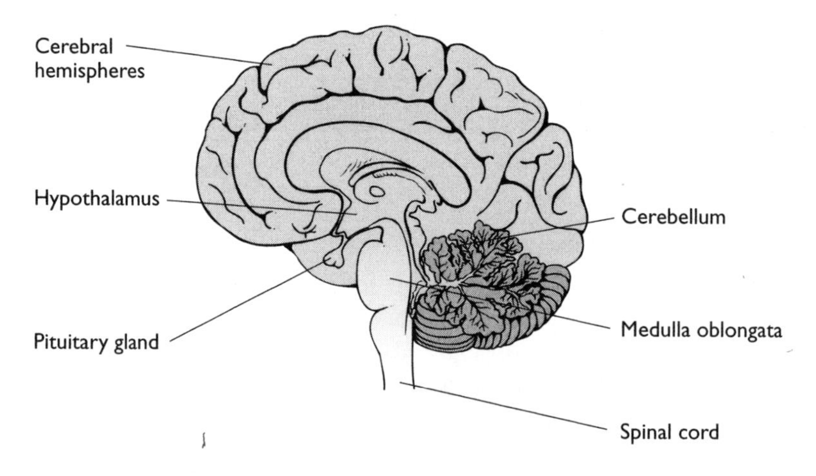

The folded surface of the cerebral hemispheres is known as the cerebral cortex. Some of the areas of the cerebral cortex are shown below.

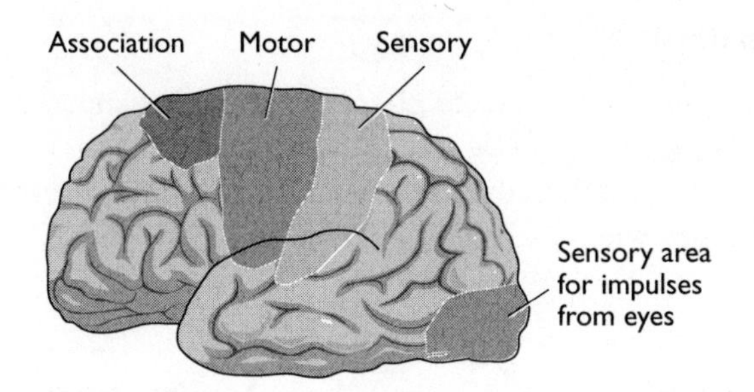

- **Sensory areas** receive input from receptors, for example in skin and muscles. Each part of the skin surface is represented in a particular part of the touch sensory area of the brain. The skin has more receptors in some parts of the body than in others, so these are represented by a greater portion of the sensory area.
- **Association areas** receive impulses from the brain's sensory areas. They then make decisions and send out impulses through the motor areas *on the opposite side of the brain*. For example, the visual association areas piece together the image and compare it with images stored in memory; they then send out nerve impulses to the necessary motor areas.

There are two areas of the cortex associated with speech:

- **Wernicke's area** is mainly concerned with understanding of both spoken and written language
- **Broca's area** is mainly concerned with controlling the muscles that produce speech

These two areas are connected by the **arcuate fasciculus**.

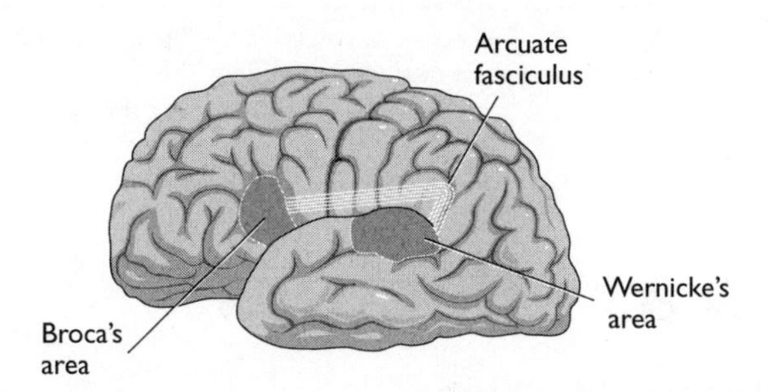

In most people these areas are in the left cerebral hemisphere.

The autonomic nervous system

The **medulla** of the brain controls cardiac muscle in the heart and smooth (unstriped) muscle. Smooth muscle is found in many parts of the body, including the blood vessels, the gut and the eyes. These muscles are involuntary.

The motor neurones controlling cardiac and smooth muscles make up the autonomic nervous system. The autonomic nervous system consists of two parts:

- the **sympathetic** nervous system
- the **parasympathetic** nervous system

The sympathetic system has opposite effects to the parasympathetic system — the two systems are antagonistic. In general, the sympathetic system enables the body to react to stress; the parasympathetic system is important in maintaining the action of involuntary muscles when the body is at rest.

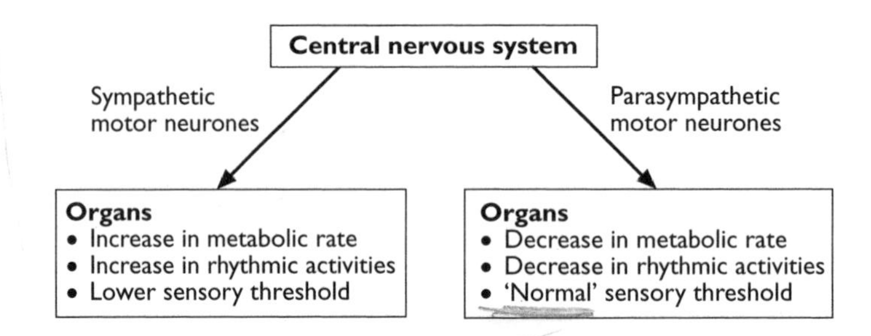

The table below shows some features of the two systems.

	Effect of the sympathetic nervous system	Effect of the parasympathetic nervous system
Neurotransmitter	Neurone endings release noradrenaline	Neurone endings release acetylcholine
Pupil	Pupil dilates	Pupil constricts
Tear production	None	Secretion of tears stimulated
Emptying of the bladder	Sphincter of bladder contracts Muscle in wall of bladder relaxes	Sphincter of bladder relaxes Muscle in wall of bladder contracts

Muscles as effectors

Antagonistic muscle action

Skeletal muscles do work by contracting — they can pull but they cannot push. Pushing movements by the body are brought about by the action of muscles on bones that act as levers.

The diagram below shows the action of some of the muscles in the leg. Some muscles occur in antagonistic pairs — they have opposite effects. One example of this is muscles A and B in the diagram, which straighten and bend the leg at the knee respectively.

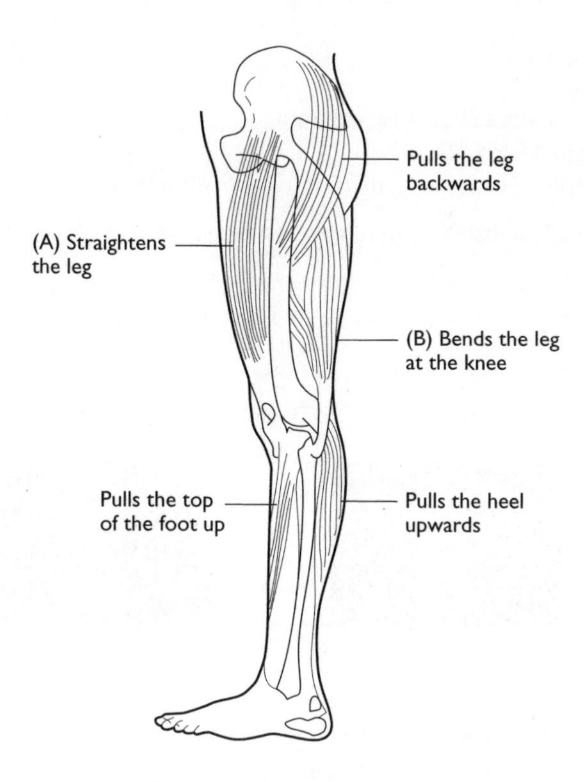

Be prepared for diagrams of muscle attachments you will not have seen before and remember that muscles can only pull on bones.

Muscle structure

Skeletal muscle is made up of striped (voluntary) muscle fibres. The photograph below shows a striped muscle fibre as seen through a light microscope.

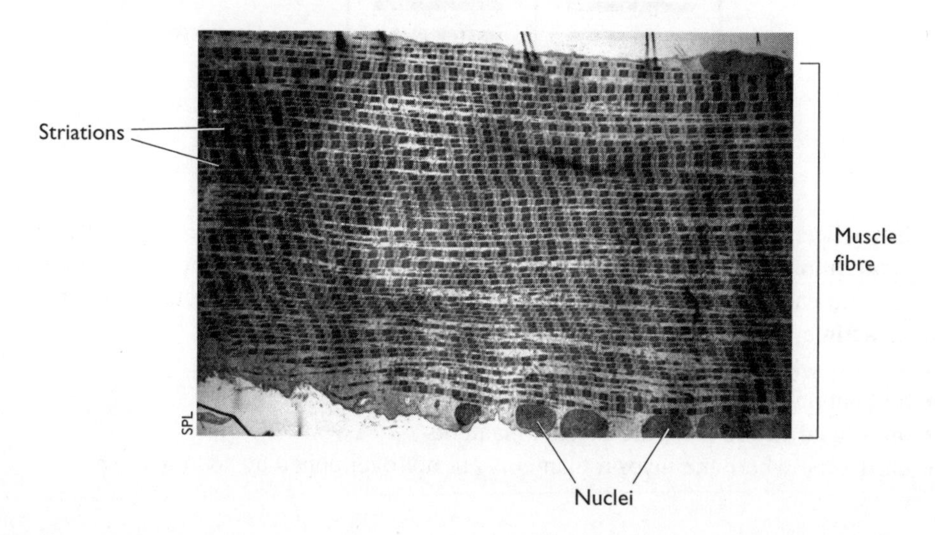

Each muscle fibre has:

- many nuclei
- an outer membrane called a **sarcolemma**
- cytoplasm called sarcoplasm
- many myofibrils, which bring about muscle contraction

Make sure that you can list the differences between a muscle fibre and a more 'typical' animal cell.

The diagram below shows the structure of a myofibril in more detail.

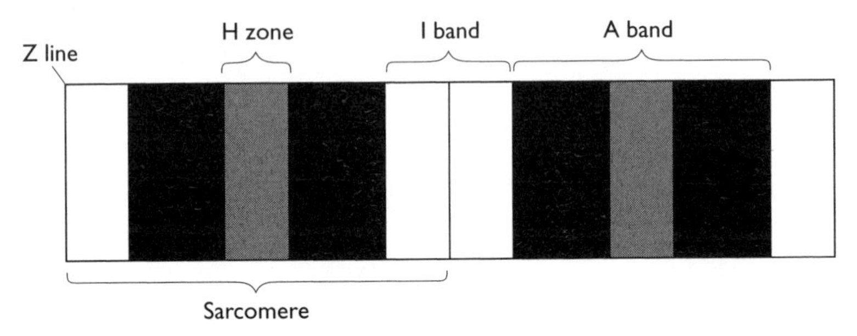

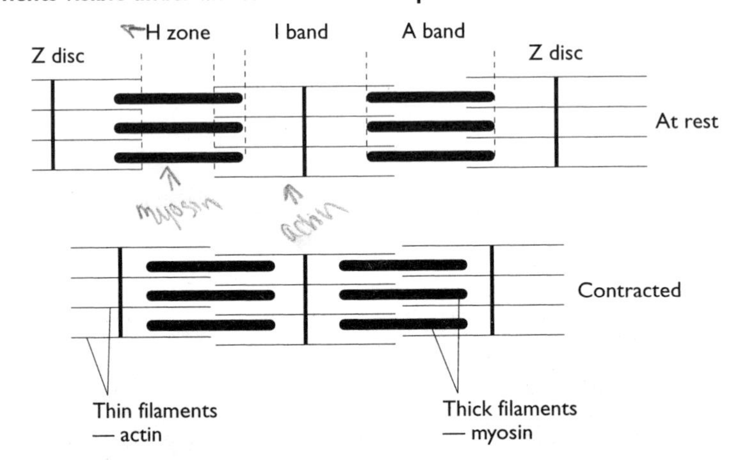

Under a powerful light microscope, repeating units called **sarcomeres**, each consisting of a pattern of light and dark bands, can be seen. Electron microscope studies have shown that each sarcomere is made up of two overlapping sets of protein filaments called **actin** and **myosin**. The regions of a sarcomere are:

- a Z disc which holds the actin filaments in position
- an I band where there are only actin filaments
- an A band where there are myosin filaments
- an H zone where the myosin filaments are not overlapped by actin filaments

Muscle contraction

Muscle contraction is explained by the **sliding filament model.**

- Cross-bridges from the myosin filaments attach to binding sites on the actin filaments.
- These cross-bridges push on the actin filaments, moving them towards the centre of the sarcomere.
- The cross-bridges disengage, then re-engage with different binding sites, acting rather like a ratchet.
- This has the effect of shortening the sarcomere, and thus the myofibril.

Take care

- The actin and myosin filaments do not contract during muscle contraction.
- The I band and the H zone both get narrower as the actin filaments are pushed towards the centre of the sarcomere.
- The A band remains the same width, as the myosin filaments do not change length.

The role of tropomyosin, calcium ions and ATP in muscle contraction

- Muscle contraction is turned on and off by a calcium switch.
- Although the actin filaments have sites where the cross-bridges from the myosin filaments can bind, these binding sites are blocked by molecules of the protein **tropomyosin**.
- When the nerve impulse arrives at the muscle fibre, calcium channels in the sarcolemma are opened; calcium ions diffuse in and bind to tropomyosin molecules, altering their shape.
- The altered tropomyosin molecules can no longer block the binding site.
- The calcium ions also activate the myosin molecules to break down ATP. This releases the energy that is used to move the cross-bridges and bring about contraction.

Inheritance

The first scientific study of inheritance was made by Mendel in the 1860s. His experiments on inheritance of characteristics in sweet peas showed that characteristics do not blend, but are passed from parents to offspring as units. These units were named genes by Johannesen in 1909. In 1912, Morgan showed that genes are located on chromosomes.

Genotype and phenotype

Genotype is the genetic constitution of an organism, i.e. all the alleles. **Phenotype** is the expression of this genotype and its interaction with the environment.

Genes and alleles

The diagram below shows the relationship between **genes**, **alleles** and **chromosomes**.

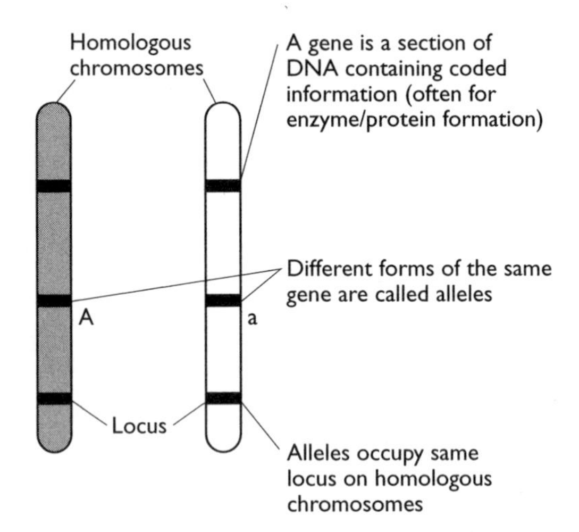

Many genes have two alleles, one **dominant** and the other **recessive**. The diagram below shows the possible combinations of such alleles.

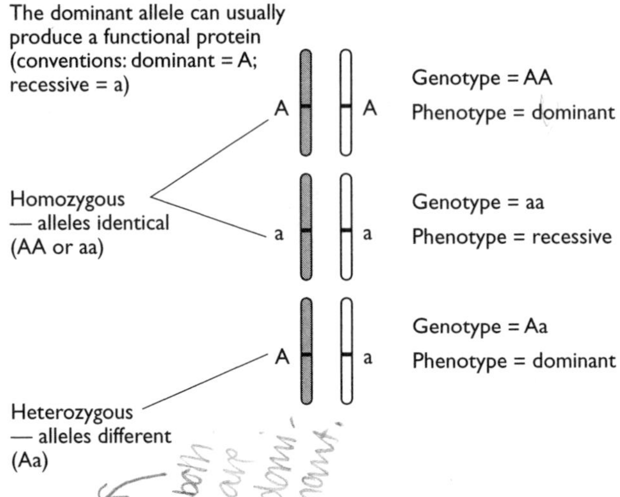

For some genes, both alleles produce a protein that can function. Such alleles are said to be **codominant**. The heterozygote in such cases will have a phenotype that exhibits features of both alleles.

Some genes have several alleles, for example the gene for ABO blood group has three alleles A, B and O.

Meiosis and fertilisation

The diagram below shows the main features of meiosis.

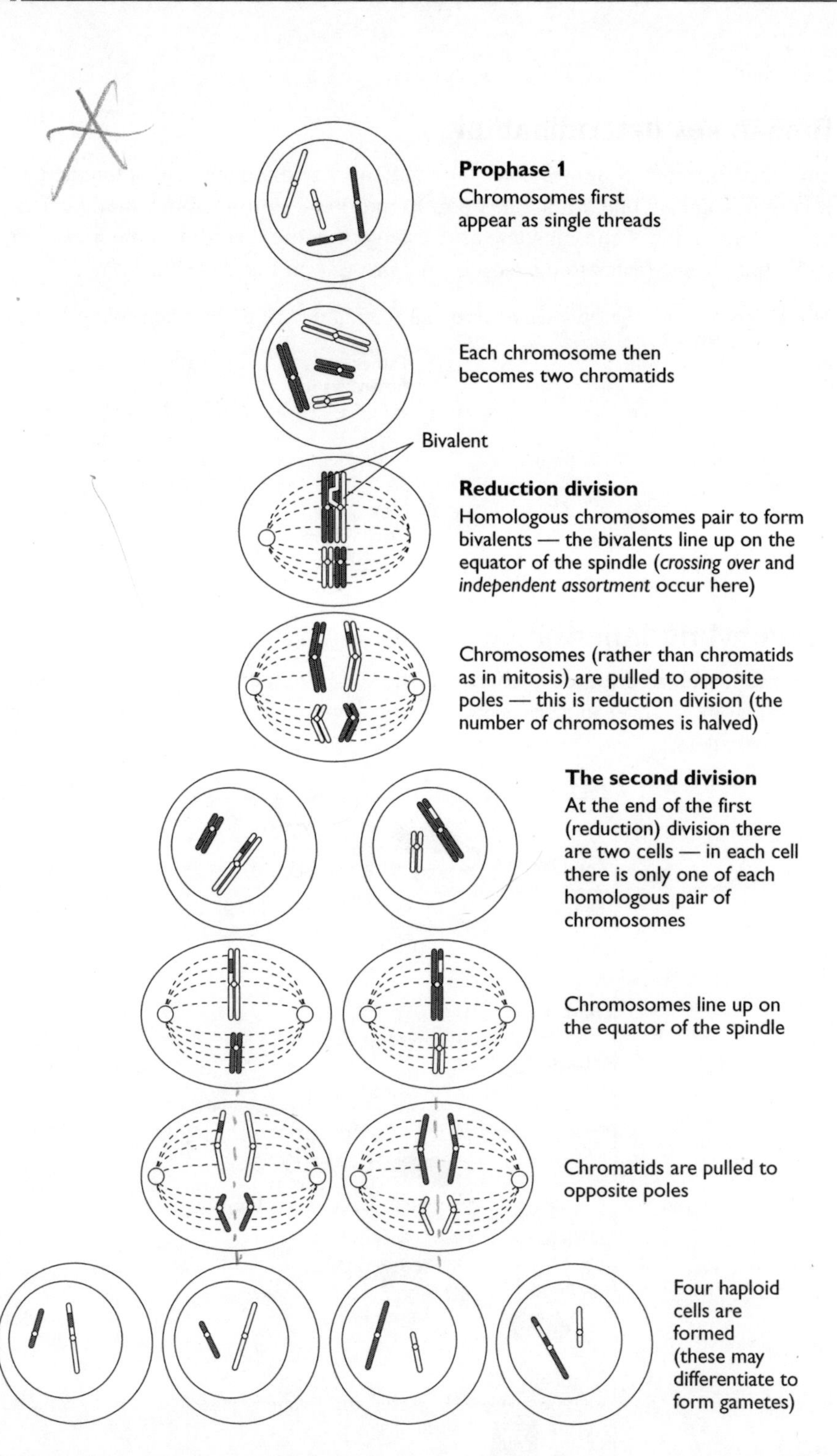

When haploid gametes fuse during fertilisation, the diploid number of chromosomes is restored.

Human sex determination

Sex chromosomes (X and Y) determine maleness and femaleness. A female has two X chromosomes while a male has one X and one Y chromosome. Many genes that are present on the X chromosome are missing from the Y chromosome. However, the gene that causes testes to develop is only present on the Y chromosome.

The diagram below shows why there is a 0.5 chance of a child being a girl.

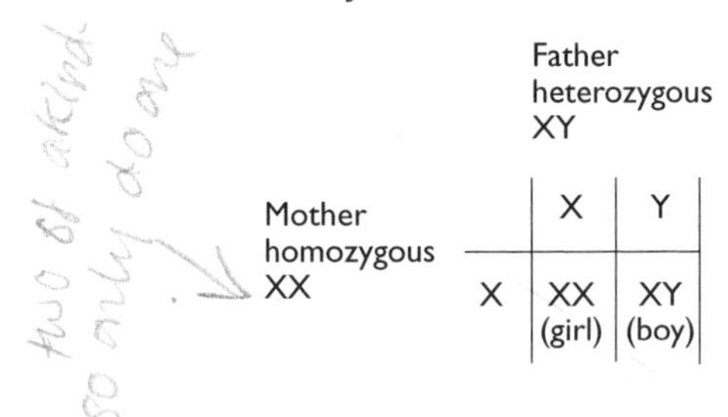

Monohybrid inheritance

In **monohybrid inheritance**, a characteristic is controlled by one gene. The following diagrams show three possible crosses for a gene which has a dominant and a recessive allele.

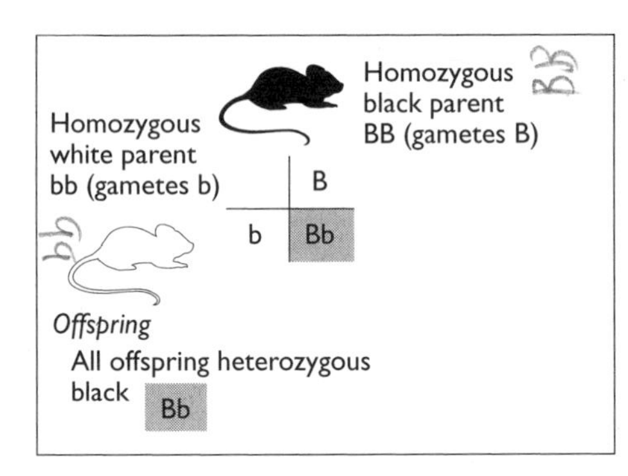

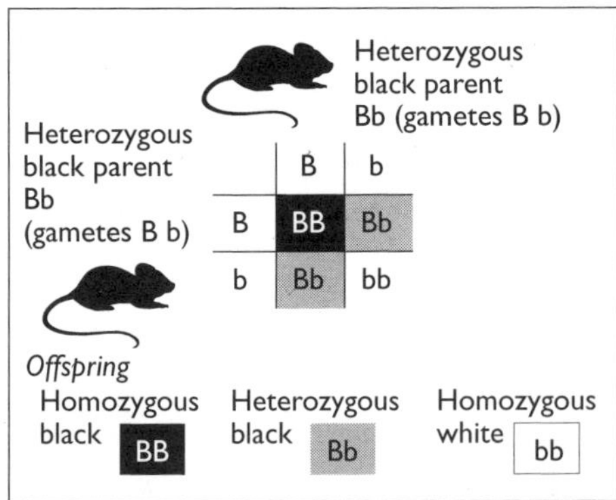

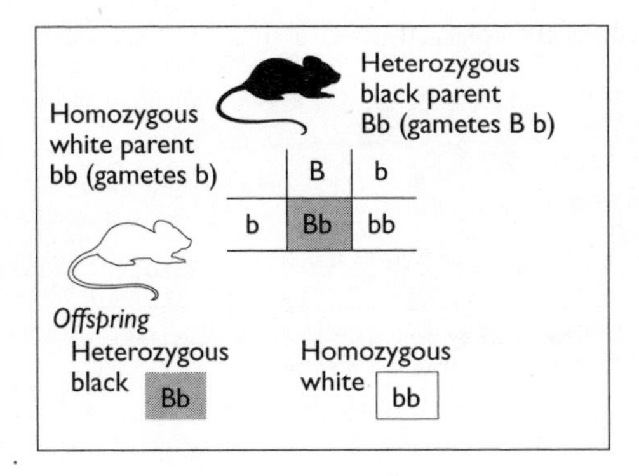

In codominance, both alleles produce a protein that can function. You must use the following conventions:

- gene C
- alleles C^A C^B
- homozygotes C^A C^A or C^B C^B
- heterozygote C^A C^B (phenotype different from either homozygote)

The diagrams below show examples of inheritance of codominant alleles.

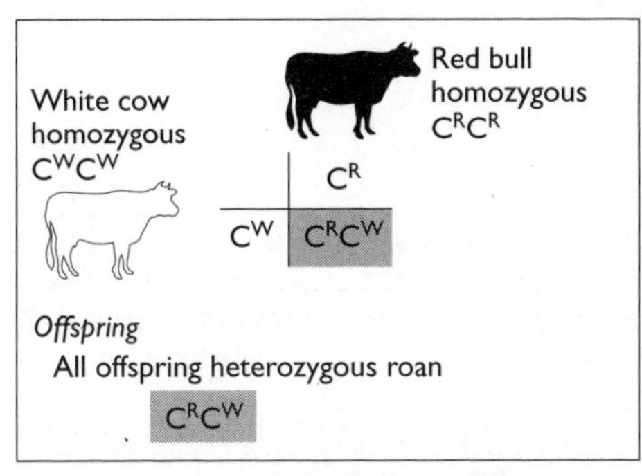

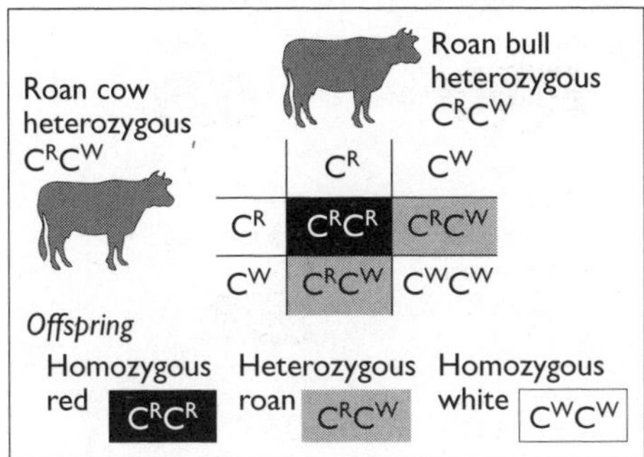

Sex-linked characters

Allele(s) for sex-linked characters are located on sex chromosomes. In humans, most alleles are not present on the Y chromosome.

You must use the following conventions for a gene with a dominant allele A, and a recessive allele a:

- homozygous female — $X^A X^A$ or $X^a X^a$
- heterozygous female — $X^A X^a$
- male — $X^A Y$ or $X^a Y$

The diagram below shows the inheritance of one form of rickets in humans.

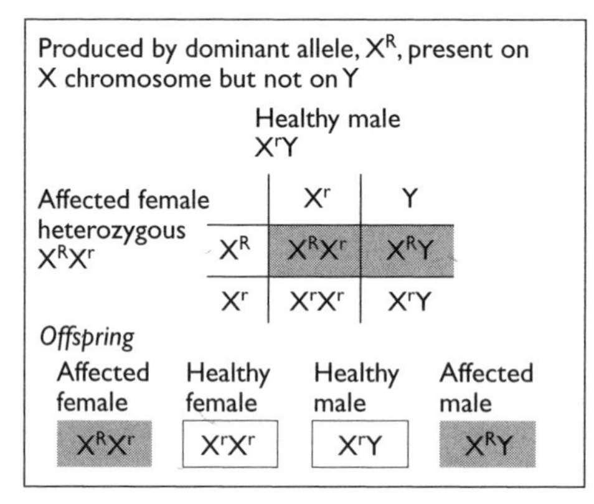

Repeat this diagram for X^rX^r crossed with X^RY.

You might also meet questions on codominant sex-linked alleles. The diagram below shows one example of this.

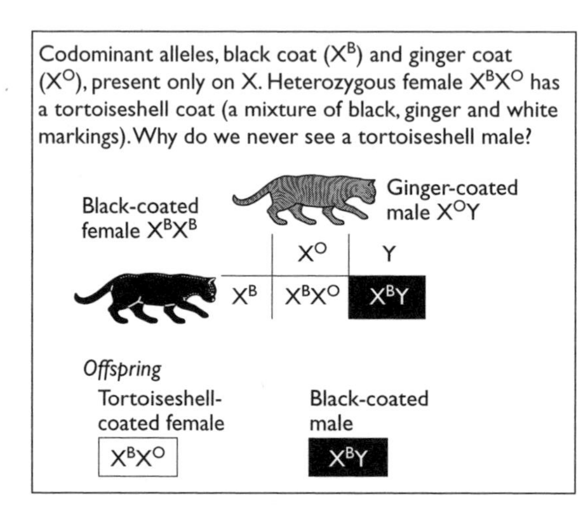

Repeat this diagram for X^OX^O crossed with X^BY.

Dihybrid inheritance

In dihybrid inheritance, a characteristic is controlled by two genes. Genes that are situated on non-homologous chromosomes are inherited independently, as shown in the diagram below.

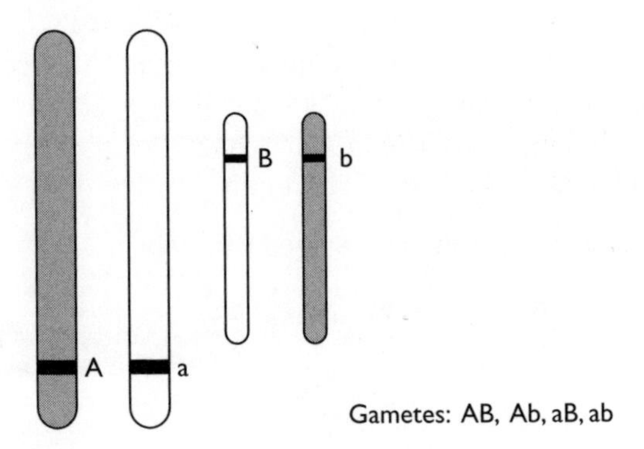

The essential principle to remember is that a gamete contains *one* allele from *each* pair, as shown in the following table.

Genotype	Alleles in gametes
AABB	AB
AAbb	Ab
AABb	AB Ab
aaBB	aB
aabb	ab
aaBb	aB ab
AaBB	AB aB
Aabb	Ab ab
AaBb	AB Ab aB ab

Notice that only the last genotype in the table, AaBb, produces four types of gamete. This means that we only get a 9:3:3:1 ratio of phenotypes when *both* parents have the genotype AaBb.

	AB	**Ab**	**aB**	**ab**
AB	AABB	AABb	AaBB	AaBb
Ab	AABb	AAbb	AaBb	Aabb
aB	AaBB	AaBb	aaBB	aaBb
ab	AaBb	Aabb	aaBb	aabb

9 offspring with both A and B alleles: AABB, 2 AABb, 2 AaBB, 4 AaBb
3 offspring with A and bb: AAbb, 2 Aabb
3 offspring with aa and B: aaBB, 2 aaBb
1 offspring with aa and bb: aabb

Ratio 1:1:1:1

The ratio can be 1:1:1:1. For example, two genes (B and L) that control colour and length of hair in hamsters are inherited independently:

- B — allele for black hair; b — allele for white hair
- L — allele for long hair; l — allele for short hair

A 1:1:1:1 ratio is achieved when BbLl is crossed with blbl:

	BL	**Bl**	**bL**	**bl**
bl	BbLl	Bbll	bbLl	bbll

Other ratios

Other ratios are also possible. Try these two crosses:

- BbLl crossed with BBll
- bbLL crossed with Bbll

Epistasis

Epistasis occurs when two genes affect a characteristic and interact with each other. One example is the control of flower colour in sweet peas. This is controlled by two pairs of alleles, A/a and B/b. The diagram below shows how these two alleles affect flower colour. Purple flowers are only produced if dominant alleles A and B are both present.

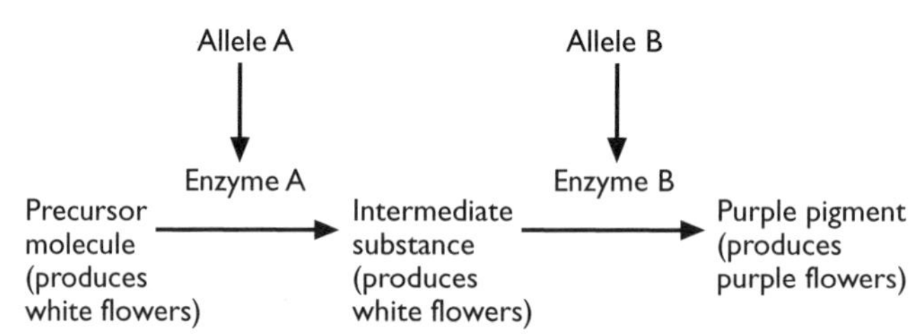

The table below shows how the genotype of the plant affects flower colour.

Genotype	Precursor substance present	Intermediate substance formed	Purple pigment formed	Flower colour
AABB	Yes	Yes	Yes	Purple
AAbb	Yes	Yes	No	White
aaBB	Yes	No	No	White
AaBb	Yes	Yes	Yes	Purple
Aabb	Yes	Yes	No	White
aaBb	Yes	No	No	White
aabb	Yes	No	No	White

Variation

Types of variation

There are two types of variation — **continuous** and **discontinuous**.

In continuous variation:

- it is not possible to distinguish distinct phenotypes
- the character is often controlled by many genes
- the character is also affected by environment

Examples include human mass and height.

In discontinuous variation:

- there are distinct phenotypes
- the character is often controlled by a single gene (which may have several alleles)

Examples include ability to taste PTC and human blood groups.

You should be able to recognise which type of variation is operating from data provided. In the graphs below, characteristic A shows continuous variation and characteristic B discontinuous variation.

Graph A
Distribution of ear length in a long-eared variety of corn

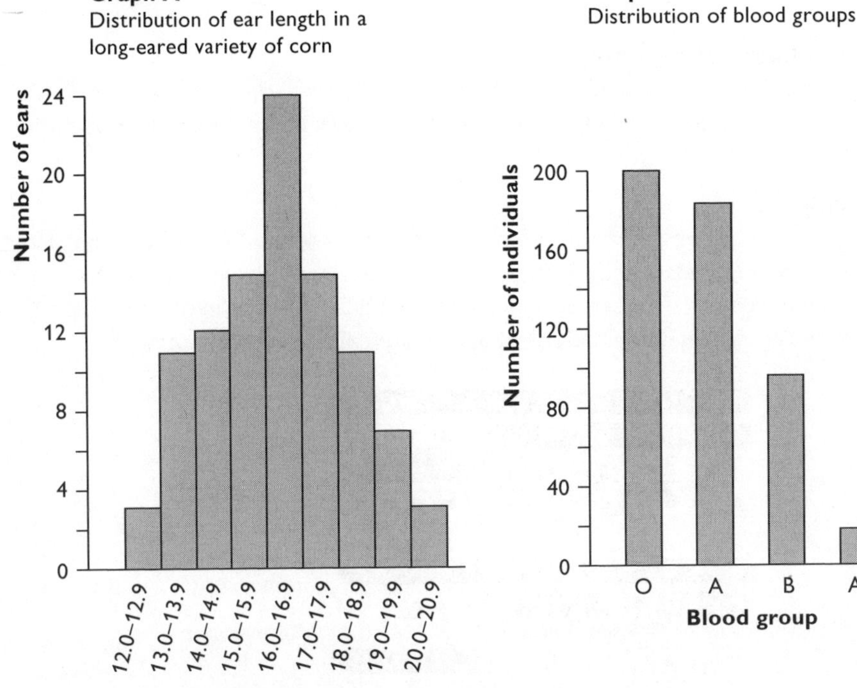

Graph B
Distribution of blood groups

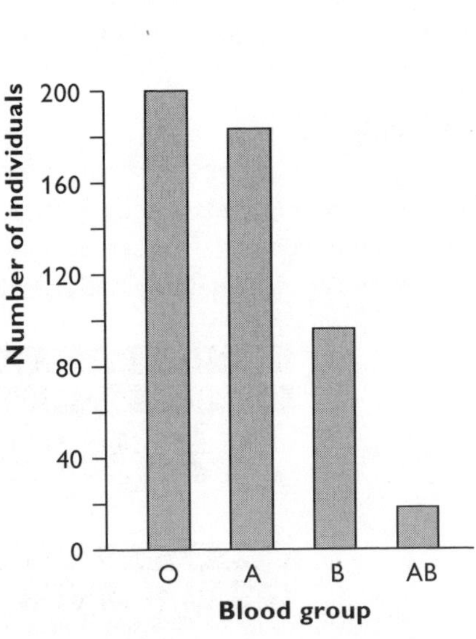

Causes of variation

Independent assortment of chromosomes

When the **bivalents** (pairs of chromosomes) line up on the equator of the spindle in the first division of meiosis, their arrangement is random. The diagram below shows two possible arrangements of two pairs of maternal and paternal chromosomes. The gametes contain different combinations of maternal and paternal chromosomes, and therefore different combinations of alleles. As the number of chromosomes increases, so does the number of ways the bivalents can be arranged. For the 23 pairs of chromosomes in humans, there are thousands of millions of different combinations of alleles arising from **independent assortment.**

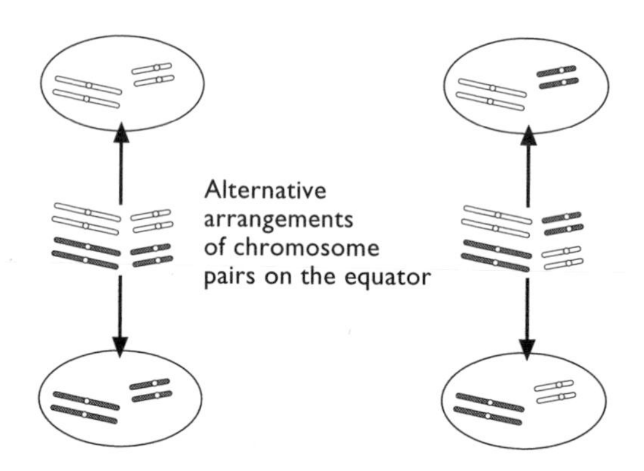

Random fusion of gametes

The gametes, which have different combinations of alleles resulting from independent assortment, also fuse randomly, leading to an infinitesimal number of combinations of alleles on fertilisation.

Crossing over

During the first meiotic division, homologous chromosomes pair to form bivalents. At this stage, **chiasmata** may form between homologous chromosomes. These can result in the crossing over of pieces of the chromatids, resulting in new combinations of alleles. These new combinations of alleles are called **recombinants**.

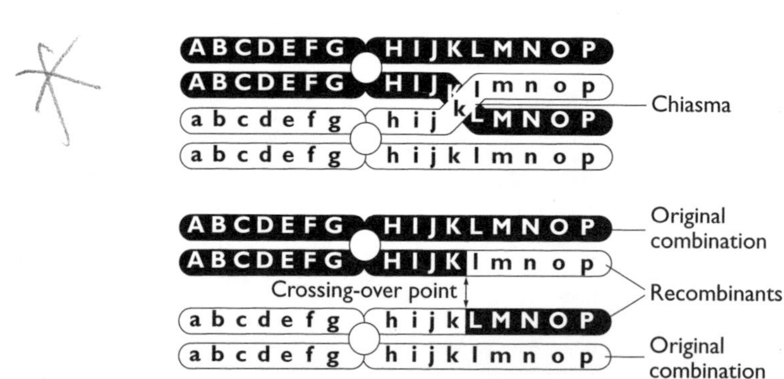

Mutation

You should revise your work on mutations from AS Unit 2. Gene mutations can result in the formation of different proteins, which in turn might affect the phenotype of an organism.

Selection and evolution

Natural selection

The modern theory of **natural selection** can be summarised as follows.

- Mutation, meiosis and random fertilisation result in genotypic variation.
- Conditions can change, for example in the environment or through the arrival of a new type of predator.
- Some changes in genotype might give rise to phenotypes that are advantageous in the new situation.
- These organisms with the new genotype are more likely to survive to breed.
- The new combinations of genes are more likely to be passed on to the next generation.
- The frequencies of the different alleles in the population will change.

You are most likely to be given examples you have not met before. You should adapt the above list to the new example by recognising the change in the organism, then explaining how this change will be advantageous and that the organisms with that genotype will be more likely to survive to breed.

Speciation

A species is a group of organisms that interbreed to produce fertile offspring and which do not normally breed with other groups of organisms. New species can arise in the following way.

- Two populations of a species become isolated.
- In each of these populations, new alleles might be formed by mutation.
- In each of these populations, combinations of alleles that enable the organisms to survive in their particular environment are selected.
- This leads to changes in the frequencies of alleles in the two populations and to phenotypic variation between the two populations.
- The populations become so different that they no longer interbreed.
- They are now regarded as different species.

The culmination of the above processes (operating on a wide variety of organisms, in a wide variety of habitats, separated from one another over hundreds of millions of years) has resulted in a great diversity of forms among living organisms.

Classification

Principles of taxonomy

The classification system used by modern biologists has two main features:

- it is a hierarchy, which means that each group is contained within a larger group and there is no overlapping between groups
- it is phylogenetic, which means that the groups are based on evolutionary relationships

The diagram below illustrates both of these points.

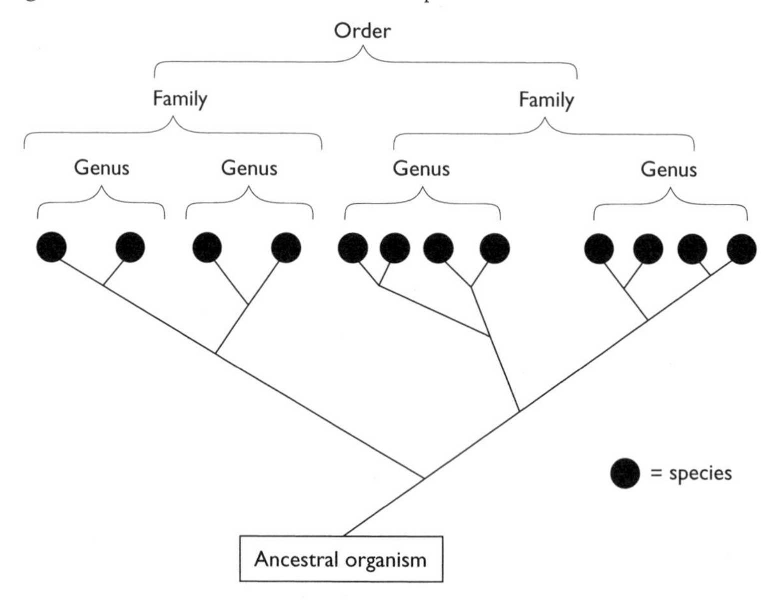

- The more recently two organisms became different species, the more closely related they are. So species that have very similar features are placed in the same genus.
- A group of closely related genera are placed in the same family.
- A group of closely related families are placed in the same order.

The full sequence, with the largest grouping first, is: **kingdom**, **phylum**, **class**, **order**, **family**, **genus**, **species**.

The five kingdoms

There are five kingdoms, each containing phyla, classes, orders, families, genera and species.

The table below lists the five kingdoms and the characteristic features of each. You should revise your AS work on prokaryotic and eukaryotic cells before learning this table.

Kingdom	Features	Examples
Prokaryotes	• Microscopic organisms with prokaryotic cells (no nuclei, mitochondria or chloroplasts)	Bacteria Blue-green bacteria
Protoctists	• Simple organisms with eukaryotic cells • They include all the organisms that cannot be placed in any of the other four kingdoms	Protozoa Algae
Fungi	• Body made up of thread-like hyphae • Cells have eukaryotic organelles • All are saprophytes or parasites • Some are unicellular	Mushrooms Moulds Yeasts
Plants	• Multicellular organisms with eukaryotic cells • Young cells have cellulose cell walls • Some cells have chloroplasts where photosynthesis occurs	Mosses Ferns Flowering plants
Animals	• Multicellular organisms with eukaryotic cells • Cells do not have cell walls or chloroplasts	Arthropods Molluscs Vertebrates

You do not need to know the classification of any organism, but you do need to know the hierarchy. Two examples are given in the table below.

	Human	Dog rose
Kingdom	Animals	Plants
Phylum	Chordates	Angiosperms
Class	Mammals	Dicotyledons
Order	Primates	Rosales
Family	Hominids	Roses
Genus	*Homo*	*Rosa*
Species	*sapiens*	*canina*

Questions & Answers

In this section of the guide there are 12 questions based on the topic areas outlined in the Content Guidance section. The section is structured as follows:

- sample questions in the style of the module
- example candidate responses mostly at the C/D boundary, though sometimes as low as grade E (Candidate A) — these answers demonstrate some strengths but many weaknesses, with potential for improvement
- example candidate responses at the A/B boundary (Candidate B) — these answers demonstrate thorough knowledge, a good understanding and an ability to deal with the data that are presented in the questions. There is, however, still some room for improvement

Some parts of the questions simply ask you to recall basic facts. Other parts contain material with which you are unfamiliar. Before answering these, ask yourself 'Which biological principle is this addressing?' Write down the principle (in rough) and then work out how the principle applies to the data. In calculations, always show your working, and when reading graphs, always draw lines between the plot in question and the axes.

Examiner's comments

All candidate responses are followed by examiner's comments. These are preceded by the icon *e* and indicate where credit is due. In the weaker answers, they also point out areas for improvement, specific problems and common errors such as lack of clarity, weak or non-existent development, irrelevance, misinterpretation of the question and mistaken meanings of words.

Question 1

Energy supply

The diagram below shows some of the relationships between respiration and photosynthesis.

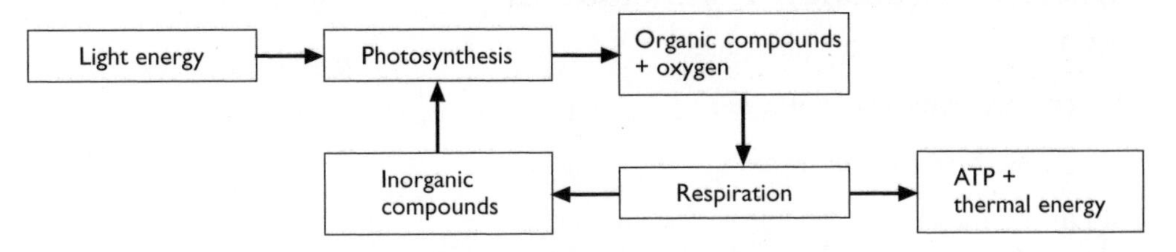

(a) (i) Name one organic compound produced by photosynthesis. (1 mark)

(ii) Name the two inorganic compounds produced by respiration. (2 marks)

(b) ATP is a source of energy for biological processes.

(i) Give two different biological processes that require energy to be transferred from ATP. (2 marks)

(ii) Explain why ATP is a useful source of energy for biological processes. (3 marks)

Total: 8 marks

■ ■ ■

Answer to question 1: candidate A

(a) (i) Carbohydrate

Carbohydrate is not a compound but a type of compound, so this response receives no mark.

(ii) Carbon dioxide and heat

Carbon dioxide is correct, for 1 mark, but heat is not a substance, and so gains no credit.

(b) (i) Taking in mineral salts, movement

Neither answer goes quite far enough to gain full credit, but together they are just about worth 1 mark. To gain the mark for mineral salts uptake, the candidate should have added '**against a concentration gradient**'. 'Movement' is rather vague; a specific term such as **muscle contraction** would guarantee the mark.

(ii) It gives up its energy readily when needed. It is then quickly formed again from respiration, so it can be used over and over again.

The first sentence is correct, for 1 mark. The second sentence is too vague. The candidate should have referred to the reversible reaction below to gain a second mark:

$$ADP + P_i + [energy] \rightleftharpoons ATP$$

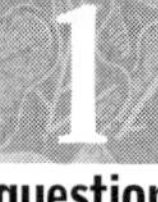

The candidate only makes two points, whereas the mark allocation is 3, so 1 mark is wasted.

■ ■ ■

Answer to question 1: candidate B

(a) (i) Glyceraldehyde 3-phosphate

(ii) Carbon dioxide and water

e All correct, for full marks.

(b) (i) Synthesis of polymers; muscle contraction

e Both correct, for 2 marks.

(ii) It can easily be broken down to release energy. This energy is released in suitable amounts. It is a small molecule, so it can diffuse easily.

e The first two sentences each contain a valid point, for 2 marks. The third sentence does not go far enough. To gain credit the candidate should have explained that this enables ATP to move from a site of production, for example a mitochondrion, to a site where energy is required, for example the cell membrane for active uptake.

Photosynthesis

(a) Oxygen is released as a by-product in the light-dependent reaction of photosynthesis. Describe the reactions that result in the formation of oxygen. (3 marks)

(b) The diagram shows part of the light-independent reaction of photosynthesis. Glyceraldehyde 3-phosphate is the first carbohydrate produced.

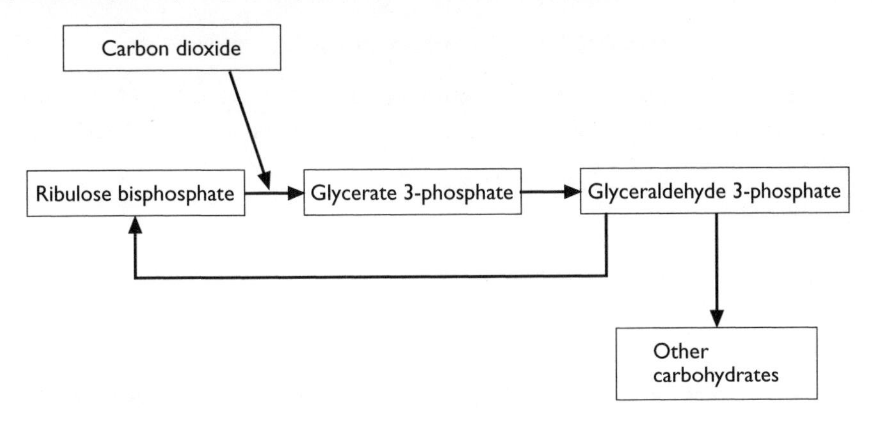

(i) How many carbon atoms has:
- **ribulose bisphosphate**
- **glycerate 3-phosphate?** (2 marks)

(ii) Name the two substances, produced in the light-dependent reaction, that are necessary for the reduction of glycerate 3-phosphate to glyceraldehyde 3-phosphate. (2 marks)

(iii) Explain what would happen if all glyceraldehyde 3-phosphate was converted to other carbohydrates. (2 marks)

(c) A group of algal cells, in an illuminated culture vessel, was supplied with carbon dioxide. The cells photosynthesised normally. After 30 minutes, the supply of carbon dioxide in the culture vessel was stopped. The graph shows the concentrations of carbon dioxide, glycerate 3-phospate and ribulose bisphosphate before and after the carbon dioxide supply was stopped.

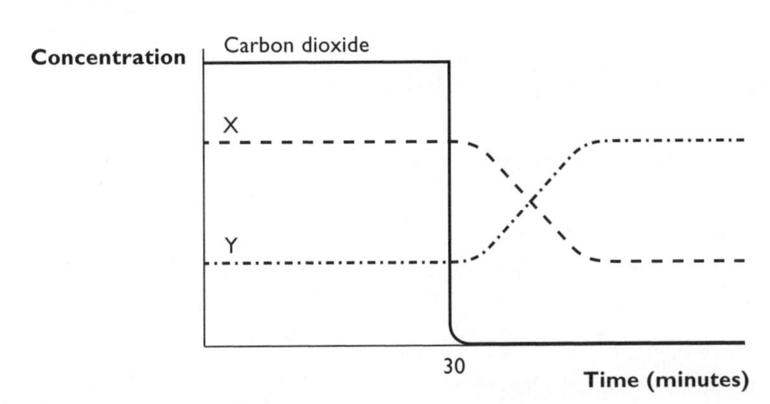

Which compound, X or Y, was glycerate 3-phosphate? Explain the reasons for your answer. (2 marks)

Total: 11 marks

■ ■ ■

Answer to question 2: candidate A

(a) Water is broken down into hydrogen atoms and oxygen atoms.

> *e* This receives 1 mark for the breakdown of water, but there is no reference to energy, and the reaction is not a simple splitting of water into hydrogen and oxygen.

(b) (i) Ribulose bisphosphate — 5C
Glycerate 3-phosphate — 3C

> *e* Both correct, for full marks.

(ii) NADP and ATP

> *e* ATP is correct for 1 mark, but the candidate should have stated **reduced** NADP for the second mark.

(iii) The process would stop.

> *e* This receives only 1 mark as it is a description of what happens rather than an explanation of why it is necessary to regenerate ribulose bisphosphate.

(c) Y is glycerate 3-phosphate. Its concentration went up because there was no NADP and ATP from the light-dependent reaction to convert the glycerate 3-phosphate to glyceraldehyde 3-phosphate, so glycerate 3-phosphate accumulated.

> *e* This deserves no marks. The candidate has not studied the graph closely and has assumed that it is a graph showing the effect of switching the light off, rather than stopping the supply of carbon dioxide.

■ ■ ■

Answer to question 2: candidate B

(a) Photolysis of water. Energy from light breaks down water into protons, electrons and oxygen.

> *e* A good answer, for 2 marks. The products are correct, but for full marks the candidate should have stated that the energy source is **excited electrons** rather than light.

(b) (i) Ribulose bisphosphate has six
Glycerate 3-phosphate has three

e The second answer is correct, for 1 mark, but in the first part the candidate has confused this reaction with glycolysis.

(ii) Reduced NAD and ATP

e ATP is correct, for 1 mark, but the candidate has given reduced NAD rather than reduced NADP, confusing the electron acceptors in photosynthesis and respiration.

(iii) The process would stop. This is because ribulose bisphosphate is the acceptor molecule for carbon dioxide. If it were not regenerated, no more carbon dioxide would be fixed.

e A good answer, for 2 marks.

(c) X was glycerate 3-phosphate. This is because its concentration went down when the carbon dioxide supply was stopped.

e This receives 1 mark for X, but the second sentence does not provide an explanation. The candidate should have gone on to state that: **glycerate 3-phosphate is only produced when carbon dioxide is fixed by ribulose bisphosphate**; and that **the glycerate 3-phosphate that had formed before the concentration of carbon dioxide was stopped had been reduced to carbohydrate**.

Respiration

The diagram shows some of the stages in aerobic respiration.

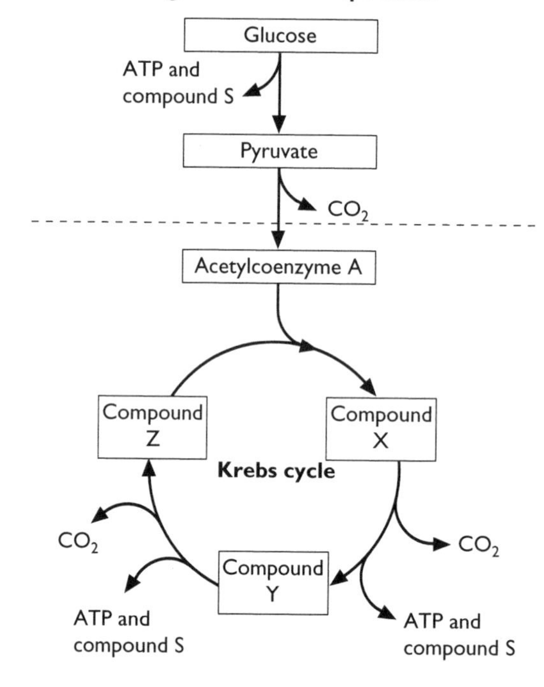

(a) (i) Name the series of reactions in which glucose is broken down into pyruvate.

(ii) In which part of the cell do these reactions take place? (2 marks)

(b) Describe the reactions that occur to form acetylcoenzyme A. (2 marks)

(c) State precisely where the reactions of the Krebs cycle take place. (2 marks)

(d) (i) Name compound S. (1 mark)

(ii) Describe what happens to compound S during aerobic respiration. (2 marks)

(e) Describe the similarities and the differences in the ways in which ATP is produced in photosynthesis and respiration. (6 marks)

Total: 15 marks

Answer to question 3: candidate A

(a) (i) Glycolysis

(ii) Mitochondria

The answer to (i) is correct, for 1 mark, but in (ii) the candidate has made the common error of assuming that all the reactions of respiration occur in the mitochondria; glycolysis occurs in the **endoplasmic reticulum**.

(b) Pyruvate reacts with coenzyme A to form acetylcoenzyme A.

This is correct as far as it goes, for 1 mark, but the candidate should have mentioned the release of carbon dioxide to gain both marks.

(c) On cristae of the mitochondria.

1 mark for mitochondria, but Krebs cycle reactions occur in the **matrix**, not on the cristae. Oxidative phosphorylation occurs on the cristae.

(d) (i) Reduced NAD

(ii) It is oxidised back to NAD.

Part (i) is correct, for 1 mark. Part (ii) receives 1 mark for oxidation, but the candidate should have mentioned **transfer of electrons to a carrier molecule** to gain the second mark.

(e) Both processes transfer energy to ADP and phosphate to produce ATP. The energy comes from light in photosynthesis but light is not involved in respiration. ATP is produced in the mitochondria in respiration but in the chloroplasts in photosynthesis.

This answer receives 3 marks for three valid points. A further 2 marks could have been gained if the candidate had given more detail, for example by stating that **sugars are oxidised in respiration** and by giving the sites of the processes in the mitochondria and chloroplasts. The sixth mark could have been gained by reference to **electron involvement** in both processes, and to the **involvement of oxygen in respiration**.

■ ■ ■

Answer to question 3: candidate B

(a) (i) Glycolysis

(ii) Outside the mitochondria, in the endoplasmic reticulum.

Both correct, for full marks.

(b) Pyruvate reacts with coenzyme A to form acetylcoenzyme A. At the same time, a molecule of carbon dioxide is released.

A comprehensive answer, worth full marks.

(c) In the matrix of the mitochondria.

Correct, for both marks.

(d) (i) Reduced NAD

(ii) It transfers electrons to electron carriers in the mitochondria. Thus it is oxidised and ready to accept more electrons from glycolysis or the Krebs cycle.

e Both correct, for full marks.

(e) In photosynthesis, electrons excited by light energy transfer energy to ADP and inorganic phosphate to produce ATP. This occurs in the grana of the chloroplasts.

In respiration, ATP is produced when electrons from reduced NAD pass down an electron transfer chain in the cristae of the mitochondria, eventually to oxygen. Reduced NAD is produced in glycolysis and the Krebs cycle.

e Everything that the candidate has written is correct, but the candidate has not answered the question, which asks for similarities and differences. The candidate is allowed half of the available marks, i.e. 3, in this situation. Simply by reorganising the material as required, this answer would gain full marks.

Survival and coordination

(a) (i) Explain what is meant by a spinal reflex. (2 marks)

(ii) Complete the diagram to show the structures involved in a spinal reflex. (3 marks)

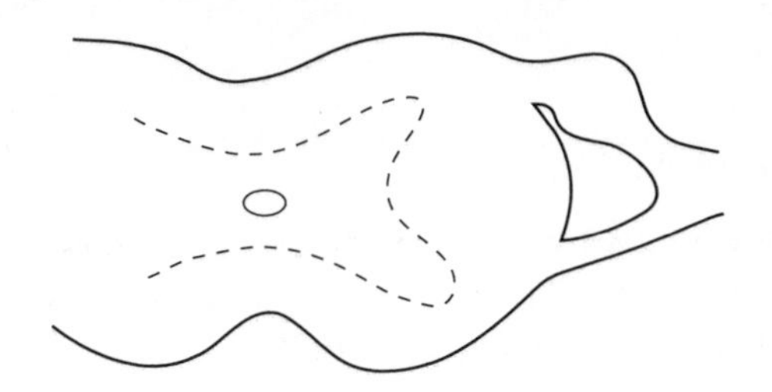

(b) The drawing shows the head of a viper. As the viper bites its prey with tooth X, liquid flows from the poison gland down through tooth X to poison the prey. For this reflex action, name the stimulus, receptor, coordinator and effector. (4 marks)

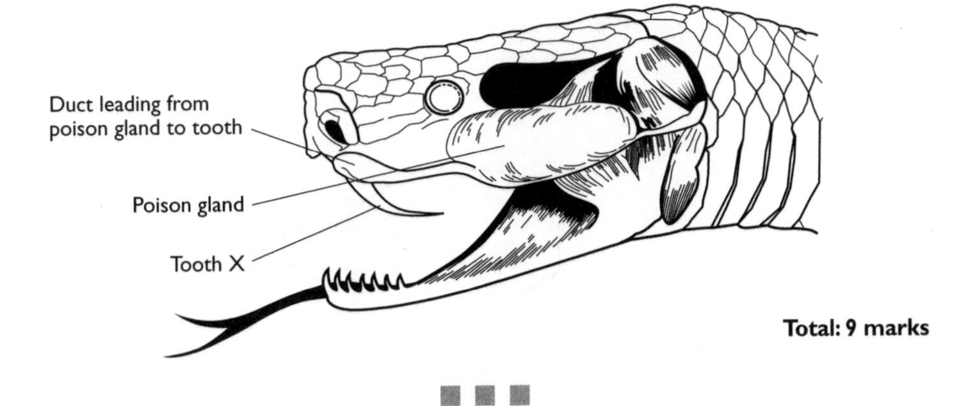

Total: 9 marks

■ ■ ■

Answer to question 4: candidate A

(a) (i) An automatic response to a stimulus.

Correct as far as it goes, for 1 mark, but 2 marks are allocated. A second mark needs a second point such as **thought or brain not involved**, or **the response is rapid**.

(ii)

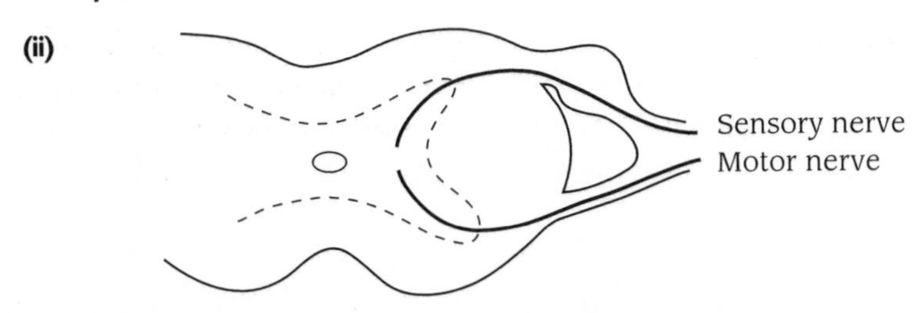

There are two major errors on this diagram. No **relay neurone** has been drawn, and the structures are labelled as nerves rather than **neurones**. However, the candidate receives 1 mark for showing that the sensory neurone enters via the dorsal root of the spinal nerve and the motor neurone leaves via the ventral root.

(b) Stimulus — prey
Receptor — tooth X
Coordinator — brain
Effector — poison gland

This answer receives 2 marks, for giving the correct coordinator and effector. The suggestions for stimulus and receptor are too vague — the stimulus is touch and the receptor is pressure receptors in tooth X.

Answer to question 4: candidate B

(a) (i) A rapid, automatic response to a stimulus.

A complete answer, for 2 marks.

(ii)

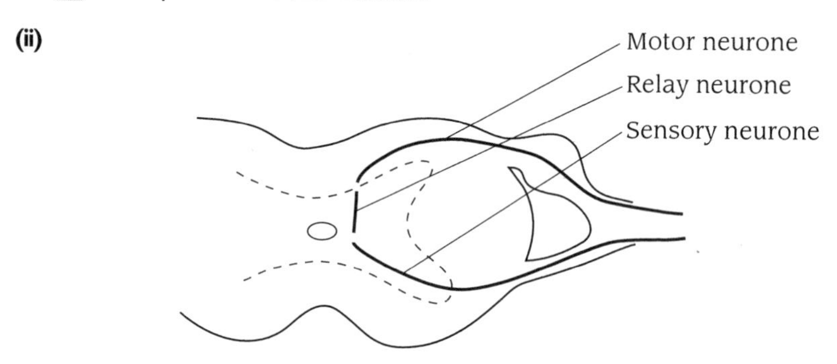

This receives 2 marks, for drawing and labelling the three neurones. However, the final mark is lost because the sensory and motor neurones are shown coming from the wrong roots of the spinal nerve.

(b) Stimulus — touching prey
Receptor — pressure receptors in teeth
Coordinator — spinal cord
Effector — poison gland

This receives 3 marks, for correctly naming the stimulus, receptor and effector. A common error is to assume that all reflex actions involve the spinal cord. When the receptor is in the **head**, the coordinator is usually the **brain**.

Homeostasis

(a) Two athletes exercised for 2 hours, and then rested for the next 24 hours. After the exercise period, one athlete was given a high-carbohydrate diet and the other athlete was given no food. The graph shows the glycogen content of the athletes' muscles over the 26 hours.

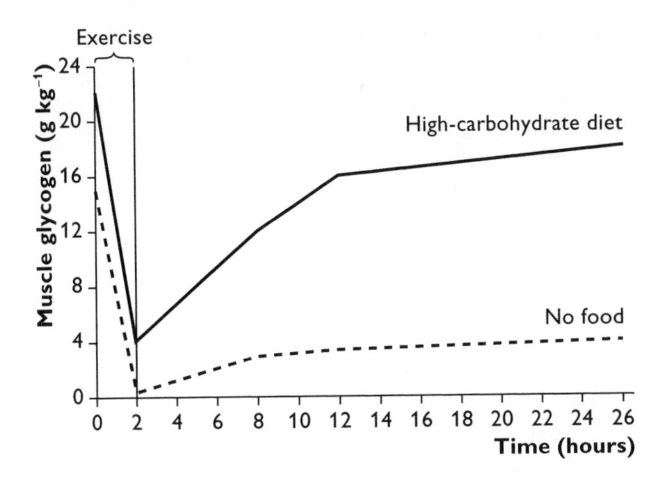

(i) Explain why the muscle glycogen concentration fell during exercise and then rose considerably in the athlete given a high-carbohydrate diet. (5 marks)

(ii) Suggest an explanation for the rise in muscle glycogen concentration that occurred in the athlete given no food. (2 marks)

(b) During the exercise period, the core body temperature of the athletes began to rise. Describe the mechanisms in the body that would operate to minimise this temperature rise. (5 marks)

(c) When we exercise, the amount of urine produced decreases. Explain why this happens. (6 marks)

Total: 18 marks

Answer to question 5: candidate A

(a) (i) The glycogen level fell because carbohydrate is needed for the energy needed to carry out exercise. When the athlete ate carbohydrate, the sugars produced when the carbohydrate was digested were converted back into glycogen in the muscles.

e This answer receives just 2 marks — for linking glycogen with energy and for linking glucose produced in digestion with increased muscle glycogen. There is no reference to **respiration**, or to the **role of insulin in increasing the uptake of glucose by the muscles**.

(ii) Glycogen in the liver was converted to glucose which was then carried to the muscles.

> This receives 1 mark as a possible suggestion, although the store of glycogen in the liver might well be exhausted during 2 hours of exercise. Fat stores are also mobilised.

(b) Capillaries in the skin would dilate, so more blood would flow nearer the surface of the skin. Sweating would increase to cool the body down.

> This receives just 2 marks — for the idea of more blood flowing near the surface of the skin, and for increased sweating. However, capillaries do *not* dilate and there is no indication as to *how* sweating cools the body. The fifth mark would be obtained by reference to the **role of the hypothalamus**.

(c) When we exercise we lose more sweat. This causes the pituitary gland to secrete ADH. ADH causes the collecting ducts in the kidneys to reabsorb more water, so less water is passed out as urine.

> This deserves 3 marks. All the statements are correct. For full marks, the candidate should have included the **stimulus** and the **receptors** for this response. There should also have been reference to the **water potential of the tissue fluid in the medulla of the kidney**.

■ ■ ■

Answer to question 5: candidate B

(a) (i) The glycogen level fell as it was converted to glucose for use in respiration to provide the energy for exercise. The carbohydrate in the diet was digested to produce glucose. When glucose was absorbed into the blood, the rise in blood glucose concentration was detected by the pancreas. This secreted insulin which converted excess glucose into glycogen.

> This is an excellent answer, for 4 marks. The only mistake is to state that insulin converts glucose into glycogen. The fifth mark could have been obtained by stating that **insulin activates enzymes that convert glucose into glycogen**, or stating that **insulin stimulates the uptake of glucose by muscle cells**.

(ii) Fat in the body could have been converted into glucose which was then converted into glycogen.

> A good suggestion, for 1 mark. The second mark could have been obtained for reference to the **role of glucagon in stimulating fat breakdown** when blood glucose levels fall.

(b) The body is cooled down by vasodilation and by the evaporation of sweat. These are brought about by the action of the thermoregulatory centre in the brain.

A reasonable answer, for 3 marks, but lacking in detail. Further credit could have been gained by referring to **increased blood supply to the skin surface, increased sweating** and **temperature receptors in the hypothalamus.**

(c) When we lose sweat, the water potential of the blood becomes more negative. This is detected by receptors in the hypothalamus which send impulses to the pituitary gland. The pituitary gland secretes ADH which is transported to the kidneys by the blood. In the kidneys, ADH makes the distal tubule and the collecting duct more permeable to water. Water passes out of the solution by osmosis due to the hypotonic solution, so less water is left in the urine.

An excellent answer, worth 5 marks, which could only have been improved by making clear *where* the hypotonic solution is. The fluid in the tubules is hypotonic and has a less negative water potential than the tissue fluid in the medulla, which has a more negative water potential and is therefore hypertonic.

Nervous coordination

(a) The diagram shows a section through the eye when it is focused on a distant object.

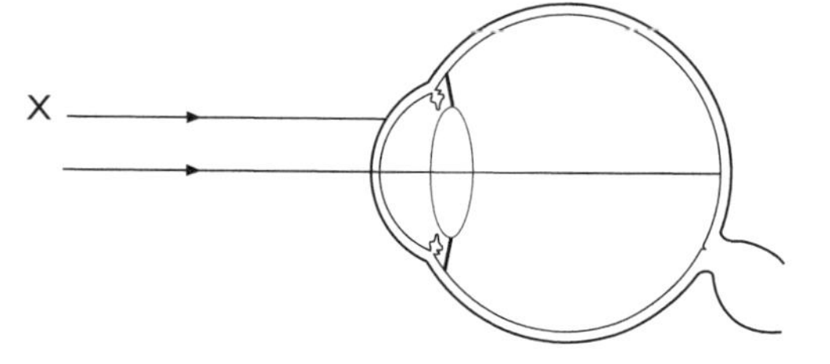

(i) Complete line X. (2 marks)

(ii) Describe the changes that occur in the eye to focus on a near object. (3 marks)

(iii) The near object is yellow. Explain how we see a yellow colour. (3 marks)

(iv) In dim light, the yellow object appears grey. Explain why. (2 marks)

(b) The diagram shows a synapse between two neurones.

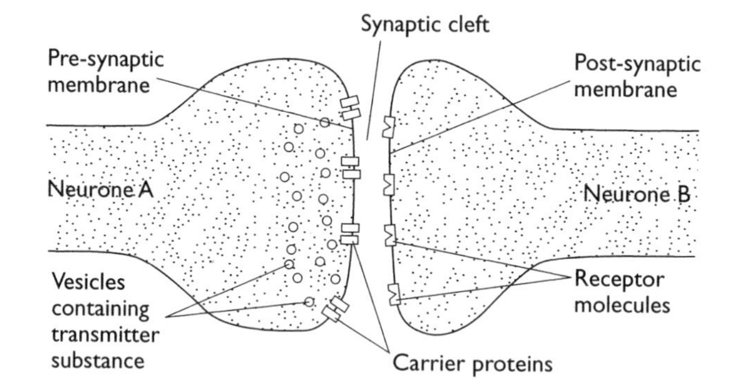

(i) Describe the events that lead to the release of the transmitter substance from neurone A into the synaptic cleft. (3 marks)

(ii) Explain how the transmitter substance causes an impulse to be initiated in neurone B. (4 marks)

(iii) Suggest the function of the carrier protein molecules in neurone A. (2 marks)

Total: 19 marks

Answer to question 6: candidate A

(a) (i)

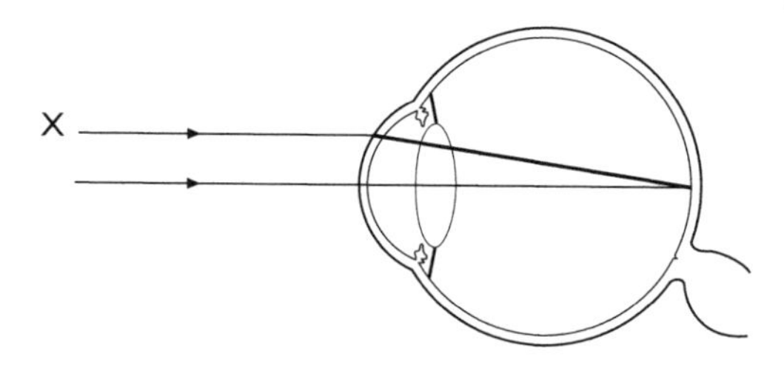

e This receives 1 mark for refraction at the cornea. For the second mark, the image should be wider than a point and, ideally, inverted.

(ii) The ciliary muscles contract and the suspensory ligaments relax. The lens then becomes fatter.

e This receives 1 mark for contraction of the ciliary muscles. The suspensory ligaments do not contract or relax. A more scientific word than 'fatter' is required — **more convex** would be ideal.

(iii) We use cones for colour vision. There are three types of cone — red, green and blue. The red and green cones decide what colour it is and send impulses to the brain.

e This receives no marks. There are two fundamental errors: we do not have different coloured cones, we have cones sensitive to different colours; and the cones cannot 'decide' anything — they transmit impulses to the brain where interpretation takes place.

(iv) The object is detected by the rods which only see in black and white.

e This receives 1 mark for the involvement of the rods. The second mark requires reference to the **difference in sensitivity between rods and cones**.

(b) (i) When the impulse reaches the pre-synaptic membrane, sodium ions enter the cell through the membrane. This causes the vesicles to move towards the membrane and release transmitter substance.

e The candidate has made the common mistake of stating that sodium ions, rather than calcium ions, enter the pre-synaptic cell. There is no reference to the fusion of the vesicles with the membrane, so this account only receives 1 mark.

(ii) It joins with protein receptors in the post-synaptic membrane. This causes the gated channels to open and sodium ions rush in, causing an action potential. The next part of the membrane becomes depolarised and the impulse passes down the post-synaptic neurone.

e A good answer, for 3 marks. The final mark would have been gained by referring to **facilitated diffusion** of the sodium ions.

(iii) Allowing sodium ions to enter.

e This suggestion receives no marks. The candidate has ignored the information that the molecules are **carrier proteins**. Sodium ions move in and out of cells via channel proteins.

■ ■ ■

question

Answer to question 6: candidate B

(a) (i)

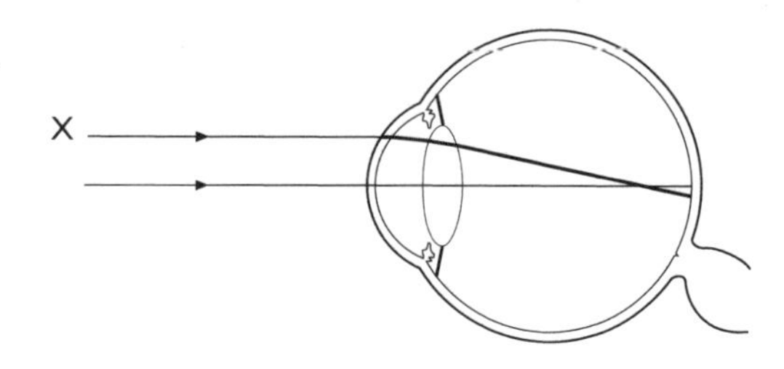

A good answer, for full marks.

(ii) The ciliary muscles contract, releasing the tension in the suspensory ligaments. The lens becomes more rounded and becomes more refractive.

A good answer, for full marks. The reference to the 'more rounded' lens does not gain credit on its own, but the additional statement about greater refraction gains the third mark.

(iii) Colours are detected by the cones. There are three types of cone, sensitive to red, green and blue light respectively. Yellow light stimulates only the red light receptors and green light receptors.

A good answer, for 2 marks. The third mark would have been scored by referring to **analysis of impulses by the brain.**

(iv) The cones have a high light threshold, so only the rods send impulses to the brain.

This receives 1 mark for the difference in sensitivity between rods and cones. For more marks there should be some reference to the fact that there is only one type of rhodopsin (as opposed to three types of iodopsin).

(b) (i) When the impulse reaches the pre-synaptic membrane, gated channels open, allowing calcium ions to enter the cell through the membrane. The vesicles move towards the membrane, fuse with it, and then release the transmitter substance.

A complete answer, for full marks.

(ii) It binds with protein receptors in the post-synaptic membrane. This causes the gated channels for sodium ions to open. Sodium diffuses in, resulting in an action potential.

A good answer, for 3 marks. The final mark would have been gained by referring to propagation of an impulse in the post-synaptic neurone.

(iii) Reabsorbing the transmitter substance so that it can be used again.

Correct, for full marks.

Analysis and integration

The drawing shows the underside of the human brain. The optic nerve from the left eye enters the brain at X.

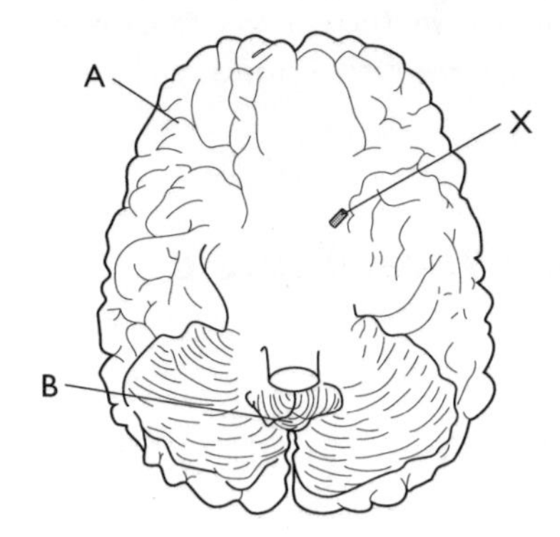

(a) (i) Name the parts labelled A and B. (2 marks)

(ii) On the diagram, use the letter Y to label the area of the brain associated with speech in most people. (2 marks)

(b) A speck of dust lands on the left eyeball. This causes tear production. Describe how the brain and the nervous system control this response. (6 marks)

Total: 10 marks

Answer to question 7: candidate A

(a) (i) A — cerebral hemisphere
B — cerebellum

A is correct, for 1 mark, but the correct answer for B is **medulla**.

(ii)

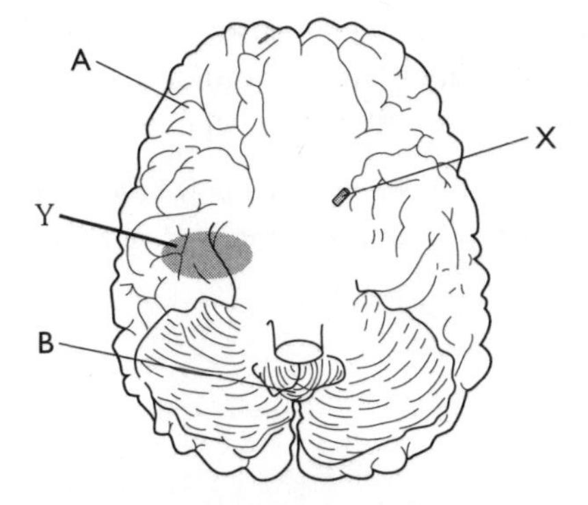

The position in the cortex is acceptable, for 1 mark, but the candidate has drawn the area on the wrong side of the brain.

(b) When the impulse enters the brain it goes first to the sensory area, then the association area. The association area sends impulses down a motor neurone to the tear gland which then secretes tears.

The candidate has not read the question correctly, and has started the answer with the brain rather than the eye. There is no reference to the **motor area** or to the **autonomic nervous system**. Only 3 marks can be awarded.

■ ■ ■

Answer to question 7: candidate B

(a) (i) A — cerebrum
B — medulla

Both correct, for full marks.

(ii)

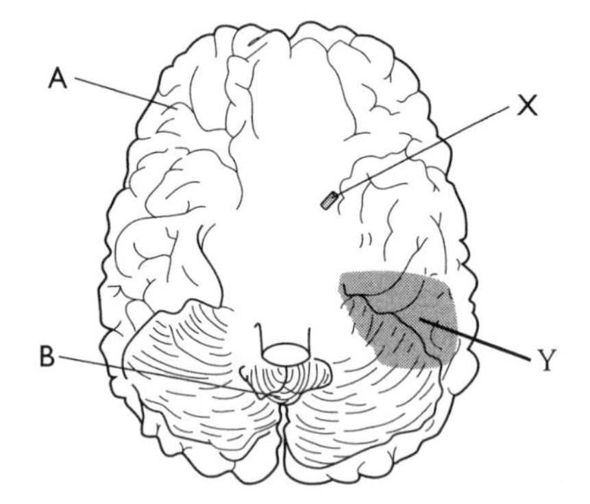

The area has been correctly drawn on the left side, but it overlaps with the cerebellum, so only 1 mark is awarded.

(b) This is an autonomic response. Touch receptors on the surface of the eye are stimulated, and impulses pass along sensory neurones in the optic nerve to the sensory cortex. The association areas coordinate the response and impulses are sent via the autonomic system to the tear glands.

A good answer for 5 marks, which could only have been improved by a clearer description of the route through the brain and reference to the **medulla** or to **motor neurones**.

Muscles as effectors

(a) The diagram shows some of the muscles in the human leg.

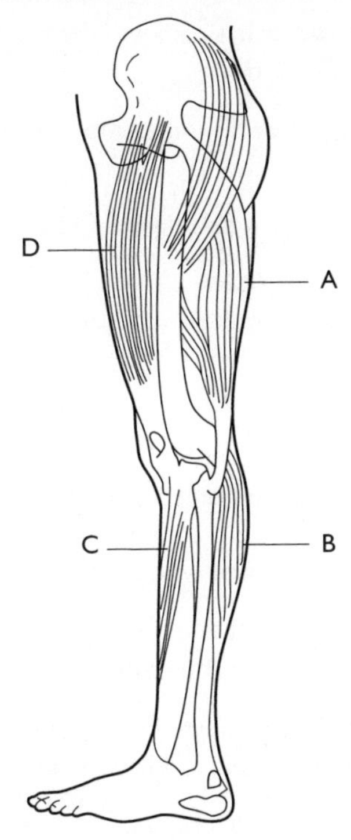

(i) Describe what happens to the leg when:

- **muscle B contracts**
- **muscle C contracts** (2 marks)

(ii) Explain why muscles such as A and D are arranged in pairs. (2 marks)

(b) The diagram shows the appearance of the sarcomeres in a striped muscle fibre.

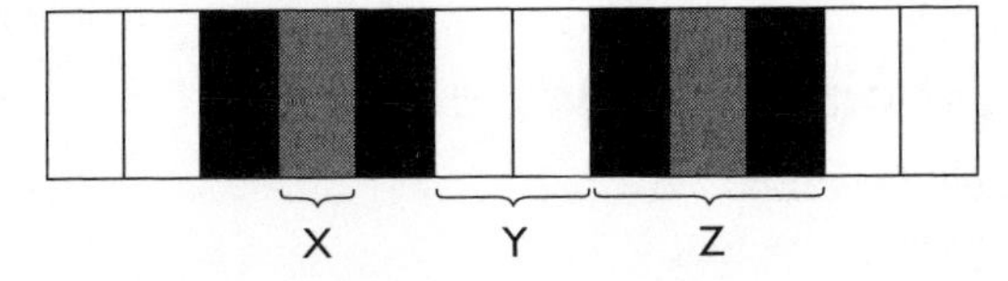

(i) Account for the appearance of the parts of the sarcomere labelled X, Y and Z. (6 marks)

(ii) Describe how and explain why the appearance of the part labelled Z changes when the sarcomere contracts. (6 marks)

(iii) Describe the role of calcium ions and tropomyosin in the contraction of a sarcomere. (3 marks)

Total: 19 marks

Answer to question 8: candidate A

(a) (i) B moves the lower leg backwards.
C pulls the foot upwards.

e In the first answer, the candidate has confused the action of muscles A and C. The second answer is correct, for 1 mark.

(ii) Two muscles are needed at every joint, one to move the bone in one direction and the other to move it back.

e This receives only 1 mark; there is no reference to **muscles only being able to do work by contracting.**

(b) (i) The muscle fibre has two kinds of filaments, called actin and myosin. The myosin filaments are darker coloured than the actin filaments, so the dark areas are myosin and the light areas are actin.

e The candidate has not referred specifically to parts X, Y and Z. There is also a fundamental error — the myosin filaments are not simply darker, they are wider. However, the candidate receives 2 marks, for knowing that there are two sets of filaments and that actin is associated with the lighter areas.

(ii) Part Z gets narrower. This is because the myosin filaments pull the actin filaments towards the middle of the band.

e This answer receives 2 marks only. There is no reference to the different behaviour of the dark and light regions of band Z when the sarcomere contracts.

(iii) Calcium and tropomyosin act as a kind of switch. When calcium ions enter the sarcomere, tropomyosin switches on contraction.

e This answer receives 2 marks. It is correct as far as it goes, but there is no detail of the relationship between tropomyosin, actin and myosin.

■ ■ ■

Answer to question 8: candidate B

(a) (i) B pulls the heel upwards.
C raises the front of the foot.

e Both correct, for full marks.

(ii) Muscles work in antagonistic pairs. When one contracts and pulls on the bone, the other relaxes. To pull the bone back to its original position the other muscle contracts.

e A good explanation, for full marks.

(b) (i) The sarcomere consists of thick myosin filaments and thinner actin filaments. Part X consists only of myosin filaments — so it looks fairly dark.

Part Y consists only of actin filaments — so it looks the lightest.
Part Z has very dark regions where action and myosin filaments overlap and a lighter region where there are only myosin filaments.

e A very good answer, for 5 marks, which could only have been improved by reference to the line in part Y. This is a disc that holds the actin filaments in position.

(ii) The dark bands stay the same width but the lighter band in the middle gets narrower. This is because the two sets of actin filaments are moved towards each other by the myosin filaments. Cross-bridges form between the two sets of filaments and it is the movement of these cross-bridges that brings about the contraction.

e A very good account for 5 marks which could only have been improved by stating that cross-bridges from the myosin filaments attach to binding sites on actin filaments.

(iii) Tropomyosin molecules block the binding sites on the actin filaments. When a nerve impulse arrives at the neuromuscular junction, calcium ions enter the muscle fibres, then pass into the sarcomeres where they change the shape of the tropomyosin molecules. These no longer block the binding sites, so muscle contraction can now take place.

e An excellent answer, for full marks.

Inheritance

Guppies are small tropical fish which are easy to breed in aquaria. Breeders have produced many different varieties. Sex in guppies is controlled in the same way as in humans. Males have the genotype XY and females XX.

The pauper variety has a red patch on its abdomen. The allele R for this colour is found only on the Y chromosome.

The dusky variety has black colouring on the rear of its abdomen. The allele B for dusky colour is dominant to the allele b for no black colouring, and is found on the X chromosome only.

Use a genetic diagram to explain the genotypes and the expected phenotypes when a female heterozygous for dusky is mated with a pauper male with no dusky colouring. (5 marks)

Total: 5 marks

Answer to question 9: candidate A

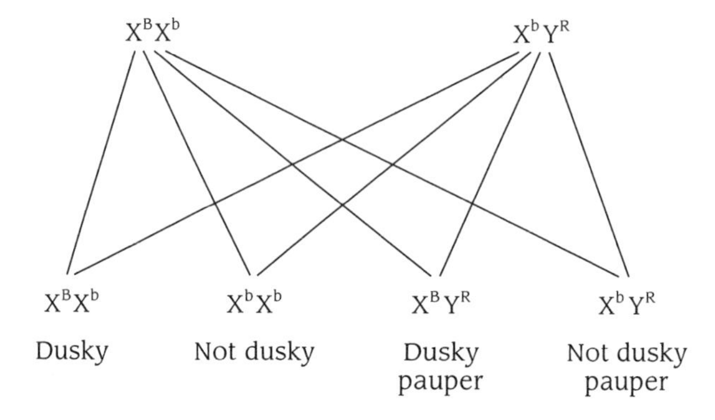

This answer receives 2 marks, for the genotypes of the parents and the genotypes of the offspring. However, the lines that the candidate has drawn do not correspond to the genotypes of the offspring, and the sex and ratio of the phenotypes are not given.

Answer to question 9: candidate B

The four genotypes are produced in a 1:1:1:1 ratio, as shown in the following Punnett square:

Male genotype
X^bY^R

Female genotype
X^BX^b

Gametes	X^b	Y^R
X^B	X^BX^b Female dusky	X^BY^R Male pauper
X^b	X^bX^b Female not dusky	X^bY^R Male pauper

This answer receives 4 marks. The candidate has correctly used a Punnett square rather than drawing lines, has identified the sex of the offspring and given the phenotype. However, the phenotype of the dusky allele has not been stated for the male offspring.

Variation

(a) The graph shows the results of an investigation into the number of petals per flower in celandine plants.

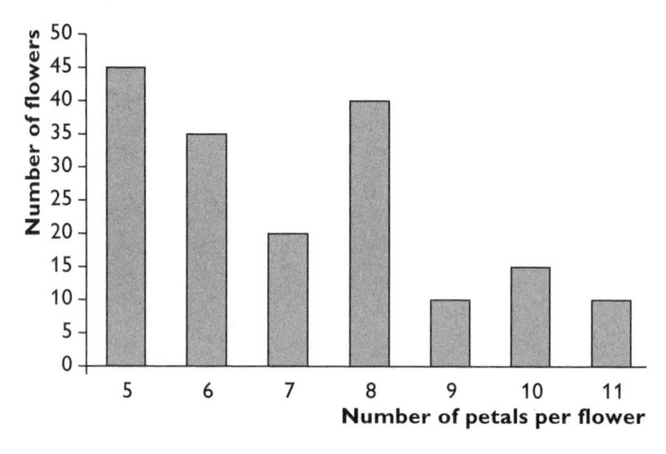

(i) What type of variation is shown by the data? Explain the reason for your answer. (2 marks)

(ii) Suggest an explanation for the variation in the number of petals per flower. (2 marks)

(b) Explain how the first division of meiosis may contribute to variation in living organisms. (5 marks)

Total: 9 marks

■ ■ ■

Answer to question 10: candidate A

(a) (i) Continuous, because there are a large number of groups.

e This answer receives no marks. Number of petals is a characteristic that is counted rather than measured; flowers have one number of petals or another. So these data show discontinuous variation.

(ii) The number of petals might be influenced by an environmental factor.

e This answer receives 1 mark, since the environment might have an effect, but there is no reference to genes.

(b) Crossing over and independent assortment both occur during the first division of meiosis.

e The candidate has correctly identified the two features that contribute to variation, for 2 marks, but has not gone on to explain *how* these features contribute.

■ ■ ■

Answer to question 10: candidate B

(a) (i) The data show discontinuous variation because each flower can be placed into a distinct class. For example, a flower has either 8 or 9 petals; it cannot have 8.7 petals.

A good answer, for full marks.

(ii) The number of petals could be influenced both by the individual's genotype and by an environmental factor.

Although candidate B has mentioned genotype, for 1 mark, there is no reference to the number of genes influencing the feature. For full marks, the candidate needed to go on to state that **discontinuous variation occurs when a characteristic is controlled by a single gene or by a small number of genes**.

(b) When bivalents form during prophase, pieces of chromatid may be exchanged between homologous chromosomes, thus leading to variation. When the bivalents come to lie on the equator of the spindle, they do so in a random way. This means that the gametes contain different combinations of chromatids.

A good answer, for 4 marks. Full marks would have been obtained if the candidate had stated that **both crossing over and independent assortment result in different combinations of alleles**.

Selection and evolution

The graph shows the estimated brain size of modern man (*Homo sapiens*) and some ancestors of humans.

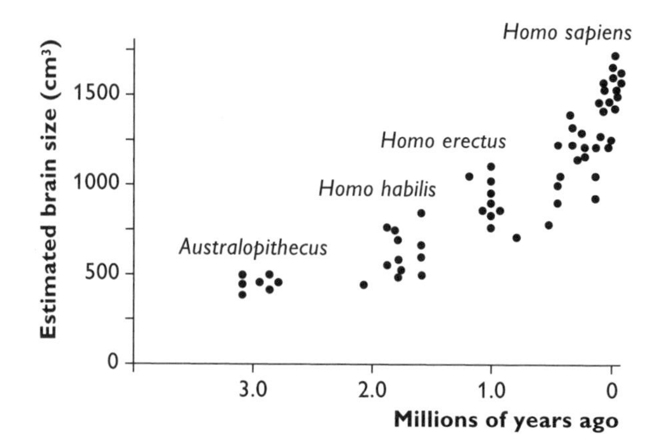

(a) Describe the pattern shown by the data. (3 marks)

(b) Suggest an explanation for this pattern. (5 marks)

Total: 8 marks

Answer to question 11: candidate A

(a) As humans evolved, their mean brain size got larger.

> The candidate has ignored the mark allocation by only giving one point, and so receives only 1 mark.

(b) Organisms with large brains were more likely to survive than organisms with small brains. They were therefore more likely to breed and their children would inherit larger brains. This is called natural selection.

> This answer receives 2 marks for the ideas that brain size is an inherited feature and that natural selection operates. There is no reference to how **variation** arose, to **genes**, or to the **selective advantage** of larger brains.

Answer to question 11: candidate B

(a) The data show that brain size increased over millions of years, and that there is a wide variation in brain size among individuals of each species.

e A good answer, for 2 marks, but there is no reference to the **exponential increase** in brain size.

(b) Mutation may have produced individuals with larger brain size. These individuals would have been more intelligent and might have had better hunting strategies. They would therefore be more likely to survive than individuals with smaller brains because they could catch more food. They would then be more likely to breed, and more likely to pass the gene for increased brain size on to the next generation. Over millions of years this led to an overall increase in brain size.

e An excellent explanation, for full marks.

Classification

(a) Explain what is meant by phylogenetic groupings. (2 marks)

(b) (i) Give the characteristic shared by all protoctists. (1 mark)

(ii) Explain why many scientists consider the protoctists to be an unsatisfactory grouping. (2 marks)

(c) Complete the table to show the classification of the lion.

	Animalia
Phylum	Chordata
	Mammalia
Order	Carnivora
	Felidae
	Panthera
Species	*leo*

(4 marks)

Total: 9 marks

■ ■ ■

Answer to question 12: candidate A

(a) Organisms are put into groups according to how closely they are related.

e This answer is good as far as it goes, for 1 mark, but neglects to mention **evolutionary history**.

(b) (i) They all have cells.

e This is too vague to gain credit.

(ii) It contains too wide a range of organisms.

e This is worth 1 mark, but states nothing about **lack of relationships**.

(c)

Kingdom	Animalia
Phylum	Chordata
Family	Mammalia
Order	Carnivora
Class	Felidae
Genus	*Panthera*
Species	*leo*

Family and Class are in the wrong places, so the candidate receives 2 marks. Learn a mnemonic to help you to remember the sequence of terms such as these. For example, **K**eep **P**lucking **C**hickens **O**r **F**ear **G**etting **S**acked.

■ ■ ■

Answer to question 12: candidate B

(a) Organisms are put into groups based on their evolutionary history. The more closely related the organisms, the smaller the group.

A good answer, for 2 marks.

(b) (i) They all have eukaryotic cells.

Correct, for 1 mark.

(ii) The group contains organisms that cannot be classified as fungi, plants or animals. It is a negative way of classifying a group rather than a positive way, based on similarities.

A good answer, for full marks.

(c)

Kingdom	Animalia
Phylum	Chordata
Class	Mammalia
Order	Carnivora
Family	Felidae
Genus	*Panthera*
Species	*leo*

Correct, for full marks.